Atmospheric Aerosols

ACS SYMPOSIUM SERIES **1005**

Atmospheric Aerosols

Characterization, Chemistry, Modeling, and Climate

Kalliat T. Valsaraj, Editor
Louisiana State University

Raghava R. Kommalapati, Editor
A&M University at Prairie View

Sponsored by the
ACS Division of Environmental Chemistry, Inc.

American Chemical Society, Washington, DC

The paper used in this publication meets the minimum requirements of American National Standard for Information Sciences—Permanence of Paper for Printed Library Materials, ANSI Z39.48–1984.

ISBN: 978-0-8412-6973-6

Copyright © 2009 American Chemical Society

Distributed by Oxford University Press

All Rights Reserved. Reprographic copying beyond that permitted by Sections 107 or 108 of the U.S. Copyright Act is allowed for internal use only, provided that a per-chapter fee of $40.25 plus $0.75 per page is paid to the Copyright Clearance Center, Inc., 222 Rosewood Drive, Danvers, MA 01923, USA. Republication or reproduction for sale of pages in this book is permitted only under license from ACS. Direct these and other permission requests to ACS Copyright Office, Publications Division, 1155 16th Street, N.W., Washington, DC 20036.

The citation of trade names and/or names of manufacturers in this publication is not to be construed as an endorsement or as approval by ACS of the commercial products or services referenced herein; nor should the mere reference herein to any drawing, specification, chemical process, or other data be regarded as a license or as a conveyance of any right or permission to the holder, reader, or any other person or corporation, to manufacture, reproduce, use, or sell any patented invention or copyrighted work that may in any way be related thereto. Registered names, trademarks, etc., used in this publication, even without specific indication thereof, are not to be considered unprotected by law.

PRINTED IN THE UNITED STATES OF AMERICA

Foreword

The ACS Symposium Series was first published in 1974 to provide a mechanism for publishing symposia quickly in book form. The purpose of the series is to publish timely, comprehensive books developed from ACS sponsored symposia based on current scientific research. Occasionally, books are developed from symposia sponsored by other organizations when the topic is of keen interest to the chemistry audience.

Before agreeing to publish a book, the proposed table of contents is reviewed for appropriate and comprehensive coverage and for interest to the audience. Some papers may be excluded to better focus the book; others may be added to provide comprehensiveness. When appropriate, overview or introductory chapters are added. Drafts of chapters are peer-reviewed prior to final acceptance or rejection, and manuscripts are prepared in camera-ready format.

As a rule, only original research papers and original review papers are included in the volumes. Verbatim reproductions of previously published papers are not accepted.

ACS Books Department

Contents

Acknowledgments ix

1. **Atmospheric Aerosols and Their Importance** 1
   Raghava R. Kommalapati and Kalliat T. Valsaraj

Aerosol Characterization

2. **The Role of Morphology on Aerosol Particle Reactivity** 13
   Eva R. Garland, Elias P. Rosen, and Tomas Baer

3. **The Characteristics and the Cytotoxic Effects of Particulate Matter in the Ambient Air of the Chiang Mai-Lamphun Basin in Northern Thailand** 31
   Narongpan Chunram, Usanaee Vinitketkumnuen, Richard L. Deming, and Richard M. Kamens

4. **Toluene Decomposition on Water Droplets in Corona Discharge** 41
   Ying Kang and Zucheng Wu

Aerosol Chemistry

5. **Surface Activity of Perfluorinated Compounds at the Air–Water Interface** 65
   Nabilah Rontu and Veronica Vaida

6. **Atmospheric Chemistry of Urban Surface Films** 79
   D. J. Donaldson, T. F. Kahan, N.-O. A. Kwamena, S. R. Handley, and C. Barbier

7. **Photochemistry of Secondary Organic Aerosol Formed from Oxidation of Monoterpenes** 91
   Stephen A. Mang, Maggie L. Walser, Xiang Pan, Jia-Hua Xing, Adam P. Bateman, Joelle S. Underwood, Anthony L. Gomez, Jiho Park, and Sergey A. Nizkorodov

8. **Effect of Highly Concentrated Dry $(NH_4)_2SO_4$ Seed Aerosols on Ozone and Secondary Organic Aerosol Formation in Aromatic Hydrocarbon/NO_x Photooxidation Systems** 111
   Zifeng Lu, Jiming Hao, Junhua Li, and Hideto Takekawa

9. **Adsorption and UV Photooxidation of Gas-Phase Phenanthrene on Atmospheric Films** 127
   Jing Chen, Franz S. Ehrenhauser, Kalliat T. Valsaraj, and Mary J. Wornat

Aerosol Modeling

10. **Understanding Climatic Effects of Aerosols: Modeling Radiative Effects of Aerosols** 149
    Tarek Ayash, Sunling Gong, and Charles Q. Jia

11. **Environmental Effects to Residential New Orleans following Hurricane Katrina: Indoor Sediment as Well as Vapor-Phase and Aerosolized Contaminants** 167
    Nicholas A. Ashley, Kalliat T. Valsaraj, and Louis J. Thibodeaux

Indexes

Author Index 185

Subject Index 187

Acknowledgments

The editors thank all those who agreed to participate in the two Symposia that we organized at the American Chemical Society Fall meeting (August 17–21, 2007) and the American Geophysical Union (May 22–25, 2007) meetings from which this Book originated. We would also thank the various authors of this Symposium Series for their timely responses to our requests regarding the manuscripts. A number of reviewers lent their time generously to provide excellent reviews on each chapter and they have to be acknowledged for their due diligence. The ACS staff who shepherded this project from the beginning have been a pleasure to work with and we thank them profusely. Professor Valsaraj acknowledges the help and support of his wife, Nisha and his two children, Viveca and Vinay during this effort. Dr. Kommalapati thanks Dr. Judy Perkins, Department Head of Civil and Environmental Engineering and Prairie View A&M University for the support that he received during this endeavor. Dr. Kommalapati takes this opportunity to thank his parents and brother for their lifetime of encouragement without which he would not be where he is today. He also thanks his wife, Suhasini and son, Rishi for their love and support. Finally, it has been an honor and privilege for Dr. Kommalapati to work on the symposiums and this book with the coeditor, Professor Valsaraj who has been his mentor and friend since his student days at Louisiana State University in the early 1990s. He expresses his sincere gratitude to Professor Valsaraj.

Kalliat T. Valsaraj
Cain Department of Chemical Engineering
Louisiana State University
Baton Rouge, LA 70803

Raghava R. Kommalapati
Department of Civil and Environmental Engineering
P.O. Box 519, Mail Stop 2510
Prairie View A&M University
Prairie View, TX 77446

Atmospheric Aerosols

Chapter 1

Atmospheric Aerosols and Their Importance

Raghava R. Kommalapati[1] and Kalliat T. Valsaraj[2]

[1]Department of Civil and Environmental Engineering, P.O. Box 519, Mail Stop 2510, Prairie View A&M University, Prairie View, TX 77446
[2]Cain Department of Chemical Engineering, Louisiana State University, Baton Rouge, LA 70803

Introduction

Apart from trace gases, the atmosphere contains a variety of liquids and solids that exist as dispersed phases in the air. They are collectively called *aerosols*. An aerosol is considered a two phase system consisting of solid or liquid particles and the gas (air) they are suspended in. Aerosols result from both natural and anthropogenic sources. Examples are dust particles generated by wind erosion of surface soils, agricultural activities, sea salt and wave breaking over oceans. Other sources include generation via chemical reactions in the atmosphere. For example, sulfate aerosols are generated by oxidation of sulfur dioxide in atmospheric moisture, particles are generated in automobile exhaust and incomplete combustion of fossil fuels in power plants. The aerosols have a direct radiative forcing effect on climate because they scatter and absorb solar and infrared radiation in the atmosphere. Aerosols also alter warm, ice and mixed-phase cloud formation processes by increasing droplet number, concentrations and ice particle concentrations. They decrease the precipitation efficiency of warm clouds and thereby cause an indirect radiative forcing associated with these changes in cloud properties. Aerosols have most likely made a significant negative contribution to the overall radiative forcing (*1*). Although the net effect is cooling, there is also evidence that black carbon in aerosols heats up the atmospheric layer in which they reside (*2*). Aerosols also impact the health of biota and has other biological effects (e.g., nutrient availability). Bioaerosols appear to have effects as cloud condensation nuclei (CCN) in some regions of the world (*3*). An important characteristic of aerosols

© 2009 American Chemical Society

is that they have varying atmospheric lifetimes. Aerosols generally have sizes ranging from 2 nm to hundreds of micrometers (4,5). Their shapes are also variable, however, an aerodynamic equivalent diameter is a useful means of representing their size. Concentration of aerosols is typically expressed in mass of particles per unit volume of the mixture (mg/m^3 or µg/m^3) or number of particles per unit volume (#/m^3) though the mass concentration is more commonly used in standards and measurements. Typical mass concentrations and particle sizes are given in Table 1 and particle sizes of various aerosol particles are listed in Table 2.

Table 1. Mass concentrations and particle sizes of aerosols (6)

Area	Concentration / µg.m^{-3}	Diameter / µm
Urban	> 100	0.03
Rural	30-50	0.07
Marine	>10	0.16

Table 2. Properties of atmospheric aerosols (7,8)

Nature of Droplet	Size/µm	Surface area/(m^2/m^3)	Liquid water content / (m^3/m^3 of air)	Typical atmospheric lifetime
Aerosols	10^{-2} -10	1x10^{-3}	$10^{-11} - 10^{-10}$	4-7 days
Fog droplets	1 – 10	8x10^{-4}	5x10^{-8}-5x10^{-7}	3 hours
Cloud Drops	10 -10^2	2x10^{-1}	10^{-7} -10^{-6}	7 hours
Raindrops	10^2 - 10^3	5x10^{-4}	10^{-7} -10^{-6}	3-15 minutes
Snowflakes	10^3 - 10^5	3x10^{-1}		15-50 minutes

The Intergovernmental Panel for Climate Change, IPCC in 2007 (1) reported significant progress over its 2001 assessment with respect to aerosol sources. The following are some of the aerosol sources along with a brief description as provided in the latest assessment report:

- Soil dust: Soil dust is a major contributor to aerosol loading and optical thickness, particularly in tropical and sub-tropical regions. Dust source regions are mainly deserts, dry lake beds, and semi-arid desert fringes, but also areas in drier regions where vegetation has been reduced or soil surfaces have been disturbed by human activities.

- Sea salt: Sea salt aerosols are generated by various physical processes, especially the bursting of entrained air bubbles during whitecap formation. This type of aerosol may be the dominant contributor to both light scattering and cloud nuclei in those regions near the marine atmosphere.
- Industrial dust, primary anthropogenic aerosols: Transportation, coal combustion, cement manufacturing, metallurgy, and waste incineration are among the industrial and technical activities that produce primary aerosol particles.
- Carbonaceous aerosols (organic and black carbon): Carbonaceous compounds make up a large but highly variable fraction of the atmospheric aerosol. Organics are the largest single component of biomass burning aerosols. The main sources for carbonaceous aerosols are biomass and fossil fuel burning, and the atmospheric oxidation of biogenic and anthropogenic volatile organic compounds (VOC).
- Primary biogenic aerosols: Primary biogenic aerosol consists of plant debris (cuticular waxes, leaf fragments, etc.), humic matter, and microbial particles (bacteria, fungi, viruses, algae, pollen, spores, etc.).
- Sulphate aerosols: Sulphate aerosols are produced by chemical reactions in the atmosphere from gaseous precursors (with the exception of sea salt sulphate and gypsum dust particles).
- Nitrate aerosols: Aerosol nitrate is closely tied to the relative abundances of ammonium and sulphate.
- Aerosols from volcanoes: Two components of volcanic emissions are of most significance for aerosols: primary dust and gaseous sulphur.

Many aerosol species (e.g., sulphates, secondary organics) are not directly emitted, but are formed in the atmosphere from gaseous precursors and aerosol species often combine to form mixed particles with optical properties and atmospheric lifetimes different from those of their components. Also clouds affect aerosols in a complex way by scavenging aerosols, by adding mass through liquid phase chemistry, and through the formation of new aerosol particles in and near clouds. The secondary aerosols which are formed as a result of atmospheric interactions also play a significant role in the atmosphere.

IPCC's fourth assessment report focus included a significant discussion on the radiative forcing due to atmospheric aerosols. Aerosols interact both directly and indirectly with the Earth's radiation budget and climate. As a direct effect, the aerosols scatter sunlight directly back into space. As an indirect effect, aerosols in the lower atmosphere can modify the size of cloud particles, changing how the clouds reflect and absorb sunlight, thereby affecting the Earth's energy budget. The indirect effects thus modify the radiative properties, amount and life time of the clouds. These direct and indirect effects of aerosols on radiative forcing are depicted schematically in Figure 1. Aerosols also can act as sites where chemical reactions take place (heterogeneous chemistry).

Aerosols also contribute to atmospheric pollution, effecting human health and visibility. For example, aerosols are implicated in mega cities pollution effects on human health. They also affect global climate in several ways. The various effects manifest themselves through the following specific interactions and between aerosols and gases in the atmosphere (6, 7):

- The condensed phase (aerosols – solid or liquid) either absorb or adsorb materials. This can lead to potential transformations of compounds and change the overall characteristics of the condensed phases. Thus, they act as catalysts for reactions in the atmosphere via heterogeneous processes.
- Solid aerosol particles can hygroscopically grow and act as solvent media for reactions such as conversion of gaseous materials to secondary organic aerosols.
- Aerosols that adsorb or absorb pollutants can mediate the wet and dry deposition of pollutants to the earth's surface and water bodies.
- Aerosols are capable of scattering light and can change the overall visibility in the lower troposphere. This can also affect the temperature and overall circulation patterns in the lower atmosphere.
- In the upper atmosphere, the presence of aerosols can significantly decrease the temperature, thus contributing to the overall greenhouse effect and global climate change.

The additional reflection caused by aerosol pollution is expected to have an effect on the climate comparable in magnitude to that of increasing concentrations of atmospheric greenhouse gases. The effect of the aerosols, however, will be opposite to the effect of the increasing atmospheric trace gases - cooling instead of warming the atmosphere. The warming effect of the greenhouse gases is expected to take place everywhere, but the cooling effect of the pollution aerosols will be somewhat regional, e.g., near and downwind of industrial areas. No one knows what the outcome will be of atmospheric warming in some regions and cooling in others. Climate models are still too primitive to provide reliable insight into the possible outcome (9).

Whereas, the effects in the lower atmosphere are better understood, there is a real lack of understanding and uncertainty with respect to the effects in the upper atmosphere. IPCC also identified that there are significant gaps in the literature on aerosol radiative forcing (Figure 2). Notice the large uncertainty in the predicted radiative forcing by aerosols and the low scientific understanding of the aerosol radiative forcing.

While there are few books available on atmospheric aerosols, they typically tend to focus on specific aspects, for example, health effects of aerosols (10) and

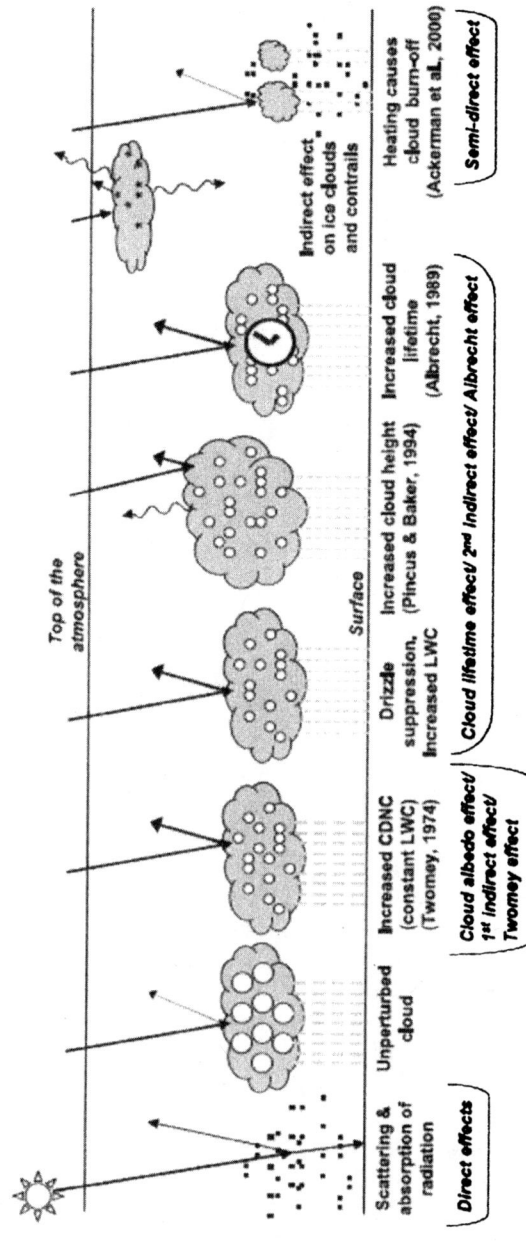

Figure 1. Schematic diagram showing the various radiative mechanisms associated with cloud effects in relation to aerosols (from Climate Change 2007, IPCC Assessment – with permission from Cambridge University Press, NY)

properties and formation aspects of aerosols (*11-13*) and aerosol chemistry (*14-16*). The present book is an attempt to address some of the gaps in the literature particularly in the area of characterization and chemistry of atmospheric aerosols and modeling. The authors of these chapters are emerging leaders in this field and their contributions presented in this book are reviews of their work along with recent advances in the field.

Organization of the Book

The concept for the book came from the two symposia that we organized at the 2007 Fall National meeting of the American Chemical Society in Boston, MA. The symposia titled "Atmospheric Aerosol Processes" attracted a good audience and the need for a specialized book on the topic was pointed out by several attendees and thus the idea for the book emerged. The book is organized into three sections: Characterization, Chemistry and Modeling of Atmospheric Aerosols.

The characterization section of the book includes three chapters. The chapters include: The role of morphology on aerosol particle reactivity; The characteristics and the cytotoxic effects of particulate matter in the ambient air of the Chiang mai-Lamphun basin in Northern Thailand; and Toluene decomposition on water droplets in corona discharge. Professor Bear's research group presented experimental evidence on the role of morphology on the reactivity of aerosol particles. The ozonolysis of oleic acid particles which are released into the atmosphere is studied and the authors concluded from their work and the evidence from other studies that the particle morphology in addition to chemical composition must be considered when evaluating particle reactivity. The chapter on fine particle matter in the ambient air presented reviews of recent investigations on the effects of air borne pollutants on human heath and specifically the levels and distribution of particulate matter (< 10 μm and <2.5 μm) along with the mutagenicity and cytotoxicity of samples collected on filters in the Chiang Mai-Lamphun basin in Thailand. Professor Wu described the toluene (a model compound representing a long lived environmental and potentially carcinogenic pollutant) decomposition on water droplets in a corona discharge. The authors propose that these kinds of reactions will have many applications in air pollution control and in industrial manufacturing processes.

The chemistry portion of the book covers several interesting topics including secondary aerosols and the chapters include: Surface activity of perfluorinated compounds at the air-water interface; Atmospheric chemistry of urban surface films; Photochemistry of secondary organic aerosol formed from oxidation of monoterpenes; Effects of highly concentrated dry $(NH_4)_2SO_4$ seed

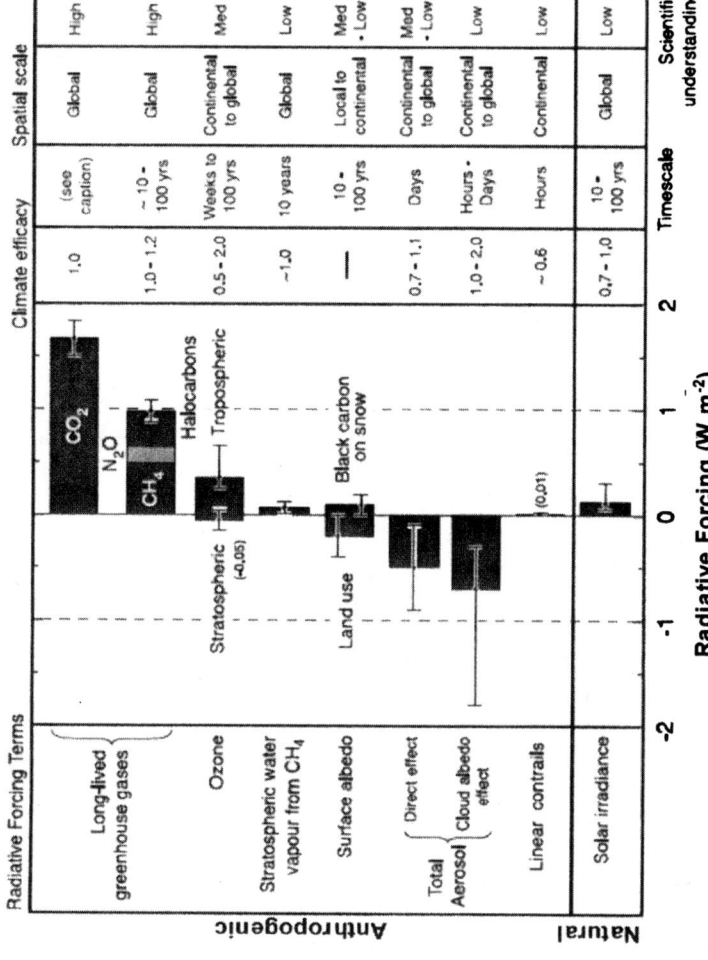

Figure 2. Anthropogenic and natural forcing of the climate for the year 2005 relative to 1750. Note the bars and vertical lines associated with aerosols. (from Climate Change 2007, IPCC Assessment - With permission from Cambridge University Press, NY).

aerosols on ozone and secondary organic aerosol formation in aromatic hydrocarbons/NO_x photooxidation Systems; and Adsorption and UV photooxidation of gas phase phenanthrene on atmospheric films. Professor Vaida and co-workers reported from experimental investigation that the perfluorinated compounds act as efficient surfactants on the surfaces of aqueous atmospheric aerosol particles and thus could potentially increase their atmospheric transport, distribution and deposition. Professor Donaldson and his group provided a summary of their on going research work on the atmospheric chemistry of large and stationary urban surface films coating buildings, roadways, etc. They hypothesize that chemistry of such films could be important in determining local oxidative strength and pollutant concentrations. The research group headed by Professor Nizkorodov in their contribution reported their work on secondary organic aerosols (SOA) which are formed from atmospheric oxidation of monoterpenes specifically the photochemical processes occurring inside these biogenic SOA particles. The photolysis of the SOA particles modifies the chemical composition of the SOA which in turn leads to emission of small volatile molecules back into the gas phase. This could explain the presence of a number of observed products in the atmosphere which are produced from the photolysis of organic peroxides and carbonyls. An international group consisting of researchers from China and Japan in their contribution presented experimental results on the effects of highly concentrated dry ammonia sulfate seed aerosols on ozone and SOA formation. The presence of highly concentrated dry aluminum sulfate aerosols had no effect on gas-phase reactions in aromatic hydrocarbon photooxidation systems, but enhanced SOA generation and SOA yield and the authors hypothesize that these results could be utilized in SOA formation modeling especially for air quality simulations involving particulate matter pollution. The last chapter in this section reports the experimental results on the adsorption and UV photooxidation of a gas phase semi volatile polycyclic aromatic hydrocarbon, phenanthrene on atmospheric water films. The interfacial partitioning of phenanthrene increased in the presence of surface active compounds . Moreover, the surface reaction of phenanthrene proceeded faster than the reaction in the bulk phase. The authors identified the different pathways of photooxidation for the surface reaction as well as reactions in the bulk water.

Finally the modeling section of the book includes two very interesting chapters; Understanding climatic effects of aerosols: modeling radiative effects of aerosols; Environmental effects to residential New Orleans following hurricane katrina: indoor sediment, vapor-phase and aerosolized contaminants. Professor Jia and his research group presented a review of the elements of aerosol-climate interactions and the uncertainties underlying aerosol-climate modeling along with the climatic implications of radiative forcing. The last chapter deals with the aerosolized contaminants in the indoor environment of residential homes in New Orleans following hurricane Katrina. The authors report that indoor pollutants are found to be more highly concentrated than contaminants found outdoors during the flood events thus making it more harmful for the residents, first responders and recovery personnel during the aftermath of Hurricane Katrina.

This book, thus brings together the varied aspects of atmospheric aerosol characterization, chemistry and modeling. It represents a unique collection of articles that will, hopefully, bring further insight into the important area of atmospheric aerosols and their importance. Further research is certainly warranted in this important area, as was indicated during the Symposium that formed the basis for this book.

References

1. IPCC, 2007, Forster et al., "Changes in Atmospheric Constituents and in Radiative Forcing", In: Climate change 2007: The Physical Science Basis, contributions of Working Group 1 to the Fourth Assessment Report of the Intergovernmental Panel on Climate Change, Cambridge University Press, Cambridge, UK, New York, NY.
2. Ramanathan, V., Ramana, M., Roberts, G., Kim, D., Corrigan, C., Cheng, C., Winkler, D., Nature, **2007**, *448*, 575-578.
3. Christner, B., Morris, C., Foreman, C., Cai, R., Sands, D., Science, **2008**, *319*, 1214.
4. Ghan J., Schwartz, S.E., 2007, Aerosol properties and processes – A path from field and laboratory measurements to global climate models, Bull. Amer. Meteor. Soc. 1059-1083, doi:10.1175/BAMS-88-7-1059.
5. Hinds W.C., 1999, Aerosol Technology: Properties, Behavior, and Measurement of Airborne Particles, 2nd ed, Wiley, New York.
6. Warneck, P., 2000, Chemistry of the Natural Atmosphere, 2nd edition, Academic Press, San Diego, CA.
7. Seinfeld, J.H., Pandis, S.N., 2006, Atmospheric Chemistry and Physics – From Air Pollution to Climate Change, 2nd edition, John Wiley & Sons, Inc., New York, NY.
8. Valsaraj, K.T., 2000, Elements of Environmental Engineering – Thermodyanmics and Kinetics, 2nd Edition, CRC Press, Boca Raton, FL.
9. NASA, 2008, "Atmospheric Aerosols", NASA Online Facts, http://oea.larc.nasa.gov/PAIS/Aerosols.html
10. Ruzer, LS., and Harley NH., 2005, Aerosols Handbook: Measurement, Dosimetry and Heath Effects, CRC Press, Boca Raton, FL.
11. Kondratyev, KY., Ivley, LS., Krapivon, VF., Varostos, CA., 2005, Atmospheric aerosol properties: Formation, processes and impacts, Springer Verlag, New York.
12. Spurny, K.R. (editor), 2000, Aerosol Chemical Processes in the Environment, Lewis Publishers, Boca Raton, FL.

13. Baumgartel, H., Grunbein, W., Hensel, F., (editors), 1999, Global Aspects of Atmospheric Chemistry, Springer, , New York, NY.
14. Meszaros, E., 1999, Fundamental of Amospheric Aerosol Chemistry, Akademiai Kiado, Budapest, Hungary.
15. Gelenscer, A. 2004, Carbonaceous Aerosols, Springer, Dordrecht, The Netherlands.
16. Colbeck, I. (editor), Environmental Chemistry of Aerosols, Blackwell Publishing Ltd., Ames, IA.

Aerosol Characterization

Chapter 2

The Role of Morphology on Aerosol Particle Reactivity

Eva R. Garland[1], Elias P. Rosen[1], and Tomas Baer[1,*]

Department of Chemistry, University of North Carolina, Chapel Hill, NC 27599

Reaction rates of aerosol particles with gas-phase species can vary by several orders of magnitude depending on the morphology of the particle. The ozonlolysis of oleic acid in mixed particles represents a system in which morphology plays a significant role in reactivity. Particularly important is the particles' surface structure, including phase and molecular arrangement. We have investigated the ozonolysis of oleic acid adsorbed onto PSL and silica substrates, and we find that the rate of ozonolysis is dependent on the composition of the core particle. Further, we find that the oleic acid does not deposit evenly onto the core particles, instead forming islands of approximately 100 nm diameter and 20 Angstroms height for our coating conditions.

Introduction

Many key atmospheric compounds, such high molecular weight polar organic compounds, partition preferentially into the condensed phase in the form of aerosol particles. In order to assess the atmospheric fate of these compounds, it is necessary to determine their reactivity under atmospherically relevant conditions. However, describing the reactivity of species in particles is significantly more complex than assessing gas-phase reactivity because the local environment of the particle can strongly influence reaction rates. Moreover, particles are usually complex mixtures of species that cannot be easily reduced to model systems.

The reactive uptake of a gas by a particle-phase reactant may be expressed in terms of a resistor model that depends on gas phase diffusion, particle phase diffusion, mass accomodation coefficient, and bulk and surface reactions within the particle (*1*). All of these terms except for gas phase diffusion can be heavily influenced by particle morphology. For example, a frozen particle would inhibit both diffusion of the condensed phase molecule to the surface of the particle and of the gas phase reactant into the bulk of the particle. Particularly important is the surface structure of the particle, since the gas phase reactant must be accomodated by the surface in order for a reaction to occur on the surface, and the gaseous reactant must penetrate the surface in order for reaction to occur in the bulk.

We focus here on the role of morphology in the ozonolysis of oleic acid particles, as this has served as a model system in the atmospheric science community for studying reactive uptake by organic aerosols. Oleic acid (cis-9-octadecenoic acid) is released into the atmosphere from meat cooking, and its dominant loss process is expected to be reaction of its double bond with ozone. Laboratory studies found that the uptake of ozone by pure oleic acid particles results in an atmospheric lifetime that is substantially shorter than observed in ambient particles (*2,3*). A possible explanation for this discrepancy is that oleic acid does not typically exist in pure particles in the atmosphere, and that other particulate components may influence the reactivity of oleic acid. In particular, oleic acid is one of many compounds that are simultaneously released into the atmosphere from meat cooking. This mixture of compounds contains several low volatility species that will preferentially partition into the condensed phase, either as mixed homogeneously nucleated particles or by absorption/adsorbtion onto preexisting particles.

Here, we examine how morphologies of internally mixed particles impact the rate of uptake of ozone by oleic acid. We first review studies on mixed oleic acid particles possessing a diverse range of morphologies. Then, we discuss recent experiments in our laboratory on the kinetis and morphology of oleic acid adsorbed onto core inorganic particles.

Internally Mixed Oleic Acid Particles

Mixtures of oleic acid with C12-C18 alkanoic acids represent a good model for studying the effect of morphology on oleic acid reactivity since saturated fatty acids and the monounsaturated oleic acid are structurally similar and miscible. Additionally, long chain saturated fatty acids are released along with oleic acid in meat cooking processes, and so investigations of these mixed particles represent a first step toward studying more atmospherically relevant systems. The phase and morphologies of mixed oleic acid/saturated fatty acid systems vary depending on the relative amounts of the components, and the microstructure of the mixture can also depend on sample preparation. Interestingly, multiple studies reveal that the reactivity of oleic acid is very sensitive to both the composition and method of preparation of the mixed particles.

Oleic Acid + Myristic Acid

The phase diagram of myristic acid (n-tetradecanoic acid) and oleic acid (Figure 1) indicates that at room temperature, the mixture is liquid for concentrations of myristic acid below $X_{MA} \sim 0.1$, and at higher concentrations, solid myristic acid exists in equilibrium with liquid myristic acid and liquid oleic acid. The right panel of Fig. 1 shows the results of studies by Nash et al. (4) on the uptake of ozone by mixed myristic acid/oleic acid particles prepared by nebulizing a solution of myristic acid and oleic acid in isopropanol. When X_{MA} exceeds 0.1, there is a dramatic drop in the uptake coefficient (γ, defined as the fraction of collisions of ozone with oleic acid that result in a reaction). At this composition, there is only a ~2% solid phase of myristic acid. In order to explain these results, Nash et al. propose that the solid myristic acid arranges itself to form a solid layer around the liquid droplet, thereby reducing the diffusion rate of ozone into the particle. This hypothesis is supported by SEM images which suggest that the particle surface is crystalline for $X_{MA} > 0.1$.

In a complimentary study, Knopf et. al (5) used a flow-tube reactor to simulate the reaction of mixed oleic acid/myristic acid particles. Their results are similar to those of Nash et al., showing that the uptake coefficient decreases significantly as X_{MA} increases to the point where solid myristic acid forms. Knopf et al. also note that the method of film preparation affected the reaction rate. Films were prepared by melting and dispersing the mixture inside a glass tube and then cooling to room temperature. Films that were cooled more slowly resulted in larger crystals, and these films reacted more slowly with ozone, possibly due to their enhanced ability to trap oleic acid in the crystals. Additionally, aged films resulted in lower uptake coefficients, perhaps due to an increase in connections of the solid network over time.

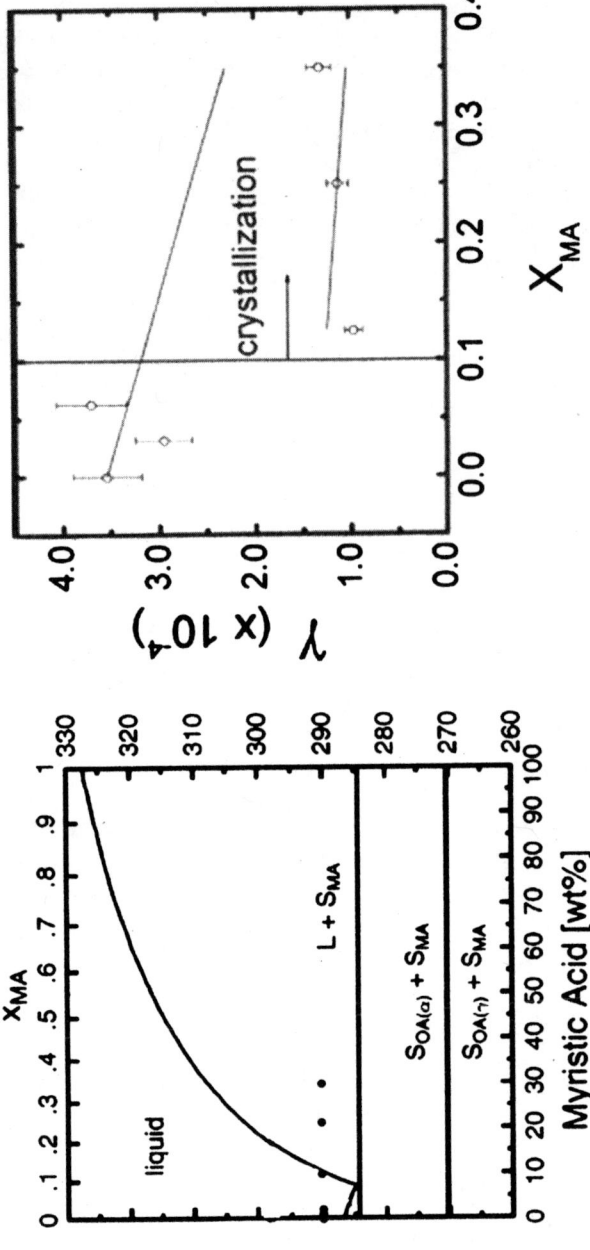

Figure 1. *Left: Phase diagram of oleic acid and myristic acid. Right: Dependence of uptake coefficient of ozone by oleic acid as a function of X_{MA}. (Reproduced from Reference 4 by permission of the PCCP Owner Societies.)*

Hearn and Smith (6) prepared mixed oleic acid/myristic acid particles by homogeneous nucleation of the vapors at temperatures of 100-150 °C, and then cooling the particles to room temperature. For $X_{MA} < 0.87$, the uptake coefficient for ozone was similar to that of the pure oleic acid particles, in contrast to the results of Nash et al. However, when the particles were further cooled prior to reaction, the uptake coefficient decreased by of a factor of 12. Hearn and Smith proposed the interesting explanation that particles were initially prepared in a supercooled (liquid) state, and further cooling allowed for crystallization to occur. Infrared analysis confirmed that the supercooled particles were liquid and that the "precooled" particles were solid. Hearn and Smith note that their results are atmospherically relevant since meat cooking is generally performed at elevated temperatures.

Other Binary Oleic Acid Mixed Systems

Katrib et al. (7) investigated the ozonolysis of mixed oleic acid/stearic acid (n-octadecanoic acid) particles. They used TEM images to observe the solidification of the stearic acid into needle-like structures. As the solid phase formed, there was a significant drop in the oleic acid ozonolysis rate, which the authors attribute to oleic acid being locked into place by the stearic acid crystal structure, and thus inaccessible to the ozone.

Ziemann (8) investigated a series of mixed particles of oleic acid with dioctyl sebacate, hexadecanoic acid, and heptadecanoic acid. The liquid oleic acid/dioctyl sebacate particles reacted at a similar rate as the pure oleic acid particles. Both the oleic acid/hexadecanoic acid and oleic acid/heptadecanoic acid paricles contained solid components. The reaction of these particles with ozone was initially fast, but then slowed significantly as the reaction progressed. Ziemann attributed these slow and fast regimes to the presence of two phases: the fast regime represents the reaction of oleic acid that is accessible to the ozone, and the slow regime represents the reaction of oleic acid that is trapped in the solid microstructure.

A detailed investigation by Hearn and Smith (9) of the oleic acid/n-docosane system revealed that even transient metastable phases can significantly affect particle morphology and reactivity. They found that after a small amount of oleic acid in the mixed particles had reacted, a metastable "rotator" phase of the n-docosane formed at the surface of the particles. This highly ordered structure decreased the rate of reaction of ozone with the oleic acid by inhibiting diffusion of ozone into the particle. When particles were first cooled to 0 °C and then reheated to room temperature, cracks formed in the surface structure, and so the particles did not display such a reduced reactivity. As the concentration of docosane increased from X(docosane)=0 to X(docosane)=0.28, the presence of the rotator phase led to a consistent decrease in reactivity with increasing X(docosane) as the surface coverage of the rotator phase of docosane increased. However, when X(docosane) exceeded 0.28, the paricles existed as an external mixture with the docosane in either the rotator or triclinic phase. Diffusion of ozone through the triclinic phase is more efficient than through the rotator phase due to cracks in the triclinic phase. Thus, the reactivity of the particles reached a minimum at X(docosane)~0.28, and increased at higher mole fractions of docosane.

Multicomponent Mixtures of Oleic Acid and Other Molecules Emitted from Meat-Cooking

While studies of binary mixed systems are important for developing a fundamental understanding of how morphological factors affect reaction rates, it is also useful to investigate more atmospherically realistic systems in order to assess whether information obtained from the binary studies can be extrapoated to atmospherically relevant situations. Both Knopf et al. (*10*) and Hearn and Smith (*11*) have studied complex mixtures of compounds that are released in meat cooking. For both of these studies, the phase of the mixture was not known, but was expected to contain both solid and liquid components.

Knopf et al. used a coated flow tube to investigate the uptake of ozone by various mixtures containing up to 15 molecules released from meat cooking. They report uptake coefficients ranging from 1.6×10^{-5} to 6.9×10^{-5}, as compared to 7.9×10^{-4} for pure oleic acid. Hearn and Smith (*11*) reacted ozone with aerosol particles released from pan frying hamburger meat. The oleic acid in these particles initially reacted quickly with the ozone, but approximately 15-30% of the oleic acid remained unreacted on the timescale of the experiment (5 sec with 100 ppm O_3). This suggests some of the oleic acid is readily accessible to ozone, while a small amount remains trapped and therefore has a longer atmospheric lifetime.

These studies, of both binary mixtures and of more complex mixtures of oleic acid with other molecules, have demonstrated the important role of morphology in the reactivity of oleic acid particles. In particular, it has been shown in several different systems that the phase of the particle can alter the uptake coefficient of ozone by more than an order of magnitude. Even a small amount of solid in the reaction matrix can effectively trap significant amounts of the liquid oleic acid and thereby slow the reaction rate.

Oleic Acid Coatings on Inorganic Particles

In addition to forming mixed particles with other organic compounds released from meat cooking, oleic acid may condense onto dry inorganic substrates (such as dust or soot) in the atmosphere. Solid particles such as soot and dust represent a large fraction of the available surface area in the boundary layer on which low volatility organic species may condense. Despite the potential atmospheric relevance, few studies have examined the adsorption of organic species onto solid inorganic particles. Katrib et al. coated oleic acid onto polystyrene latex spheres, and monitored the change in aerodynamic diameter and density of the particles as they reacted with ozone (*12*). A recent study by Kwamena and Abbatt showed that the rate of PAH oxidation by ozone is faster when the PAH is bound to azaleic acid aerosols compared to phenylsiloxane oil aerosols (*13*).

We have investigated the reactivity and morphology of oleic acid coatings on silica and polystyrene particles. We find that both the structure of the substrate and the nature of the oleic acid coating influence the rate of ozonolysis of the oleic acid.

Experimental

Sample preparation

The core particles used in this study consisted of nonporous silica particles and polystyrene latex spheres of 1.6 micron diameter (Duke Scientific). The particles were suspended in a 50/50 water/methanol solution, and atomized with a commercial atomizer (TSI). The atomized stream passed through a heated tube and a diffusion dryer to remove the solvent. For coating experiments, particles were sent through an oven containing heated oleic acid (*14*).

In order to assess quantitatively the structure of the coating, we used a combination of AFM, ellipsometry, and contact angle goniometry. Due to the requirement of the measurement techniques, these experiments were performed using flat substrate surfaces. A flat substrate is a good approximation for the micron-sized particles used in this study since the particles' radius of curvature is large compared to the size of an oleic acid molecule. The similarity in our AFM results on flat surfaces and our SEM results on particles (see Results) further support this assumption. Additionally, we found that SEM images of oleic acid coatings on the flat substrates were similar to those of the coated particles.

Flat silica substrates were prepared by treating a silicon(100) wafer with UV-O_3 in a commercial instrument followed by washing with de-ionized water. Ellipsometry measurements indicate that the native SiO_2 coating on the silicon wafer was ~10^{-15} Å thick after this cleaning procedure. A flat polystyrene surface was prepared by spin-coating polystyrene dissolved in toluene on a silica substrate followed by annealing for 12-24 hours at 130 °C (*15*).

Oleic acid was deposited on the silica and polystyrene surfaces via vapor deposition in an oven at 70 °C. The oleic acid vapor pressure at 70 °C is 3-4 x 10^{-5} Torr.(*14,16*) The oleic acid was pre-heated for several minutes, and then the substrate was placed upside down on top of a beaker containing a reservoir of liquid oleic acid. A second larger inverted beaker covered the setup in order to contain the oleic acid vapor.

Kinetics Studies with Aerosol Time of Flight Mass Spectrometry

Particle-bound oleic acid and ozone reacted in a flow tube, and the oleic acid was monitored using an Aerosol Time-of-Flight Mass Spectrometer (ATOFMS). Details of the set-up have been presented elsewhere (*4*). Briefly, coated particles were introduced into a 1m long by 1.5" ID glass flow tube. Ozone, produced by an ozone generator (Pacific Ozone Technology), was injected through a ¼" glass tube that could be moved over the length of the flow tube to vary the interaction time between the ozone and oleic acid. Flow rates of ozone and aerosol were velocity matched to ensure laminar flow through the flow tube with stable mixing.

After traversing the flow tube, the particles entered the ATOFMS through an aerodynamic lens, which focused the particles into a narrow beam. The focused aerosol then passed through the incident beams of two 532 nm diode lasers where the vacuum aerodynamic diameter of each particle was measured. Upon arrival in the center of the instrument, a particle was first volatilized by a CO_2 laser (λ=9.3-10µm) and subsequently ionized by a 118.5-nm pulsed VUV laser beam. This two-step laser desorption/ionization process was used to maximize the amount of organic coating vaporized and to minimize the fragmentation of the oleic acid. The resulting ions were focused into a 1 m long drift tube terminated by a multichannel electron multiplier for mass analysis.

The rate of reaction of oleic acid with ozone was calculated by monitoring the change in the integrated signal of the oleic acid molecular ion ($M^{+\cdot}$, m/z=282) as a function of ozone exposure, which is the product of the ozone concentration in the flow tube and the time of interaction between the ozone and aerosol. Prior to data collection, laser powers and temporal separation between IR and VUV firing were adjusted to maximize the intensity of the $M^{+\cdot}$ peak. The molecular ion signal intensity on both core types correlated positively with the temperature of the oven containing oleic acid, indicating that the volume of oleic acid on the particle increased with increasing oleic acid vapor pressure in the oven, as expected.

Surface characterization

SEM images were taken with a Hitachi S-4700 scanning electron microscope. For the images presented here, particles were sputter-coated with 2 nm of gold prior to analysis. Images were also taken without sputter-coating, and similar surface structure was observed as for the sputter-coated samples. However, charging effects decreased the quality of the non-sputter coated images.

AFM measurements were taken with a Thermomicroscopes Autoprobe M5 AFM. Tips had a force constant of 5 N/m and a reported radius of curvature of 40 nm. Image backgrounds were flattened with AutoProbe Image software. A Visual Basic Program was written to determine the number of islands and peak heights at a variety of deposition times. This program selected for islands that were a minimum of 10 Å above the background signal.

Ellipsometry data was taken with a Rudolph Research Auto EL null ellipsometer. Ellipsometry provides a measurement of the average thickness of a film based on the change in polarization of grazing incidence light, assuming the index of refraction of the coating is known. An index of refraction of 1.46 was used for all measurements. Conveniently, this is the index of refraction for both silica and for oleic acid. The index of refraction for bulk polystyrene is 1.58, and is a sharp function of orientation of the polymer chains and of film thickness (*17*). Since we were only concerned with the approximate thickness of the polystyrene layer, no effort was made to correct for its index of refraction.

Contact angle goniometry was performed by pipetting a drop of water onto a bare or coated silica surface. A picture of the water droplet on the surface was taken, and the angle between the surface and the droplet was measured with image analysis software.

Kinetics Results

Extensive kinetic analysis of pure oleic acid ozonolysis indicates that the reaction takes place predominately at or near the surface of the particle. Under the formalism of the resistor model for this case, the concentration of oleic acid decays exponentially with respect to its initial value (*18*):

$$\frac{[Oleic]}{[Oleic]_0} = \exp\left(-\frac{SA}{V} k_2 H P_{o,} t\right) \qquad (1)$$

where SA is the oleic acid surface area, V is the oleic acid volume, k_2 is the rate constant for reaction at the suface, H is the Henry's law solubility constant for O_3 in oleic acid, P_{O3} is the partial pressure of ozone, and t is the elapsed reaction time. All experiments were conducted with ozone in excess, so that fitting an exponential function to the observed decay of oleic acid signal intensity with ozone exposure yielded a pseudo-first-order rate constant, k, for the reaction of oleic acid and ozone on each inorganic core under the different coating conditions, where k=(SA/V) k_2H and has the units [1/atm·s].

Figure 2 shows a summary of pseudo-first-order rate constants for all oleic acid ozonolysis experiments conducted on spherical polystyrene latex and silica aerosol particles (19). The averaged rate constants are clearly stratified by inorganic core type, with the reaction on polystyrene latex proceeding approximately 40% faster than on silica particles. Reaction between ozone and the inorganic cores proceeds orders of magnitude slower than the reaction with oleic acid. Therefore, we believe that the observed difference in rate constants is due to the different chemical characteristics of the two core types. Silica, possessing terminal OH groups, is hydrophilic in nature, while the polystyrene latex is hydrophobic. These disparate polarities may result in different strengths and geometries of the oleic acid/substrate interaction on the two surfaces. In particular, self-coordination between oleic acid molecules that has been hypothesized to influence its reactivity with ozone (18), and interactions with the core particle surface may disrupt this self-coordination.

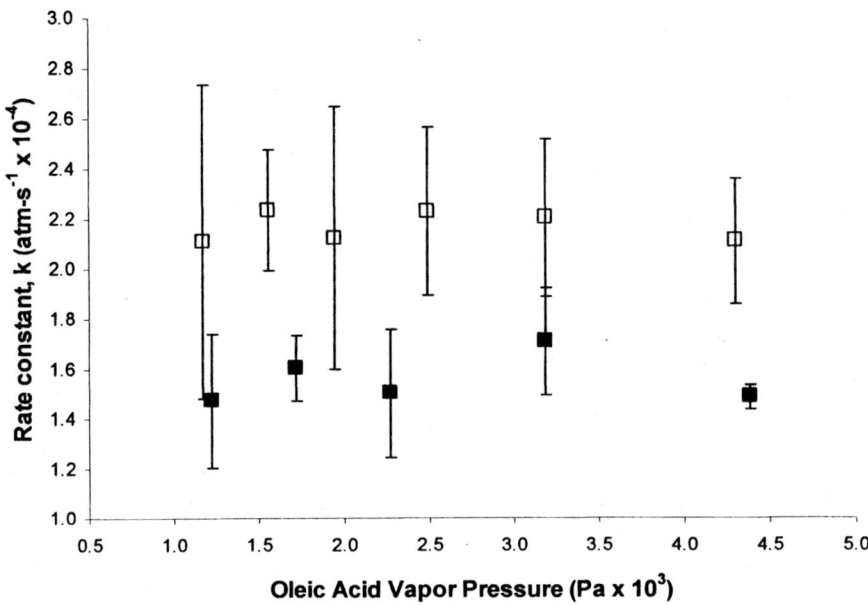

Figure 2. Psedo-first-order rate constants for oleic acid ozonolysis on coated silica (filled squares) and polystyrene latex particles (open squares). Each data point is an average of 2-6 experiments.

Interestingly, the rate constants do not change perceptibly as a function of oleic acid vapor pressure. As noted, mass spectral information indicated that more oleic acid was present on the inorganic cores as the vapor pressure of oleic acid increased in the coating oven, and Equation 1 suggests that the rate constant should vary with SA/V. Furthermore, an evenly coated particle would be expected to have a larger d_a than the core particle. However, velocimetry measurements showed that the aerosol aerodynamic diameter (d_a) did not change as the amount of coating increased. We note that d_a is related to the volume-equivalent diameter (d) by the formula $d_a = d(p_a/p\chi)^{1/2}$, where p_a is the density of the particle, p is unit density, and χ is the shape factor, defined as the ratio of resistance drag of the particle to that of a sphere having the same volume (20). χ is 1 for a spherical particle, and is greater than one for a nonspherical particle. Thus, an increase in χ would result in a lower d_a (12). An alternate explanation for a decrease in d_a with increased coating would be a decrease in p_a. Since the density of oleic acid is lower than the density of the two cores employed in this study, it is possible that a lower p_a could be contributing to our observed decrease in d_a. However, our calculations suggest that the decrease in d_a cannot be explained by a decrease in p_a alone. Although d can be calculated directly if the mobility diameter (d_m) is known, standard mobility particle sizers are limited to particles smaller than those used in our study.

Coating Morphology

To examine the possible role of the shape factor χ, and to better understand our kinetic data, we imaged both polar (silica) and nonpolar (polystyrene) particles coated with oleic acid using SEM (21). We further investigated the structure of oleic acid coated on flat surfaces of silica and polystyrene with a combination of analytical tools including AFM, ellipsometry, and contact angle goniometry. We find that oleic acid forms islands on both polar silica substrates and on nonpolar polystyrene substrates.

Oleic acid coatings on particles

SEM images of 1.6 micron PSL particles both uncoated and coated with oleic acid are shown in Figure 3. The coated particles show the oleic acid covering the surface unevenly, in the form of islands. We also took SEM images of uncoated and coated silica particles, but the rough surface of the uncoated particles made it more difficult to distinguish differences in morphology between the coated and uncoated silica particles.

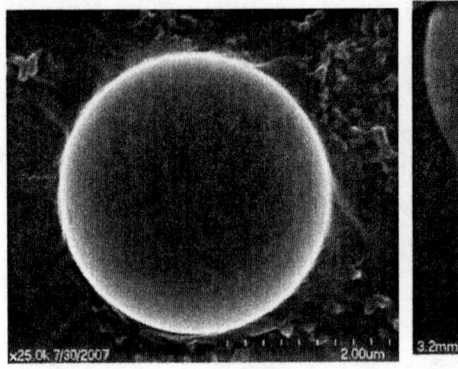 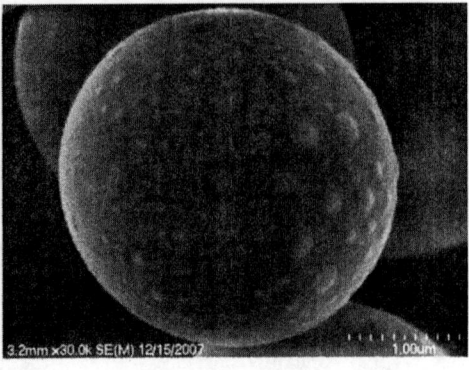

Figure 3. SEM images of 1.6 μm polystyrene latex spheres uncoated (left) and coated with oleic acid (right). (Reproduced from Reference 21 by permission of the PCCP Owner Societies.)

The SEM results motivated us to investigate further the morphology of oleic acid coatings on polar (silica) and nonpolar (polystyrene) surfaces. In order to take advantage of a variety of analytical techniques, and to better quantitate the results, we performed the remainder of the experiments using flat substrates with a surface layer of SiO_2 or polystyrene (see Experimental).

Oleic acid coatings on silica surfaces

Ellipsometry was used to measure the average thickness of the oleic acid coating as a function of deposition time on silica substrates, and the results are shown in Figure 4. Also plotted is the contact angle of a water drop on the surface, as obtained with contact angle goniometry. We note that the thickness of the oleic acid layer increases rapidly at first and then more slowly at longer times. This is consistent with a growth model reported by Kubono et al. (*22*) for films involving cluster formation. Furthermore, the contact angle measurements scale with the amount of coating, indicating that the surface is becoming more hydrophobic.

AFM images of the silica substrates exposed to different amount of oleic acid are shown in Figure 5. Both images show that oleic acid forms islands on the silica surface rather than coating the surface evenly. The diameter of these islands is approximately 100 nm, a value that is an upper limit because the resolution may be limited by the tip curvature radius.

We further examined formation of islands as a function of deposition time by comparing the number of islands and peak heights at a variety of deposition

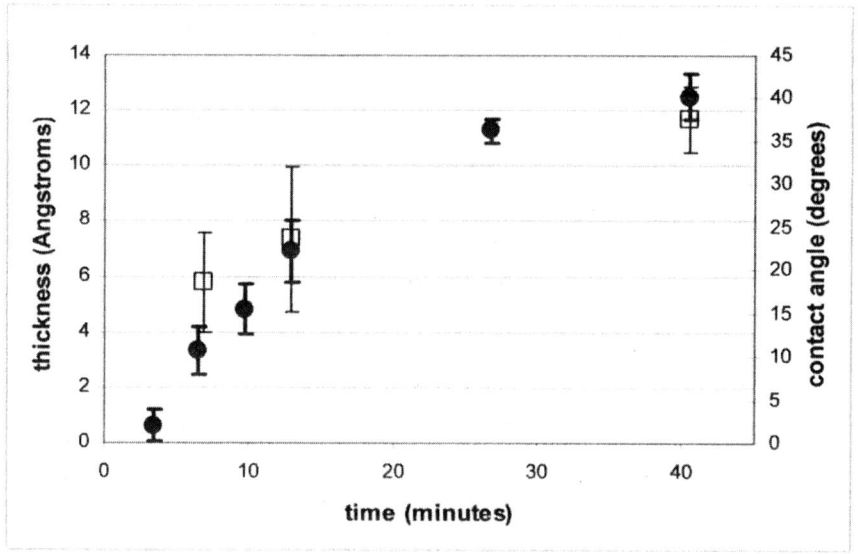

Figure 4. Deposition of oleic acid on silica measured with ellipsometry (solid circles) and contact angle goniometry (open squares). All error bars represent one standard deviation of the measurements. (Reproduced from Reference 21 by permission of the PCCP Owner Societies.)

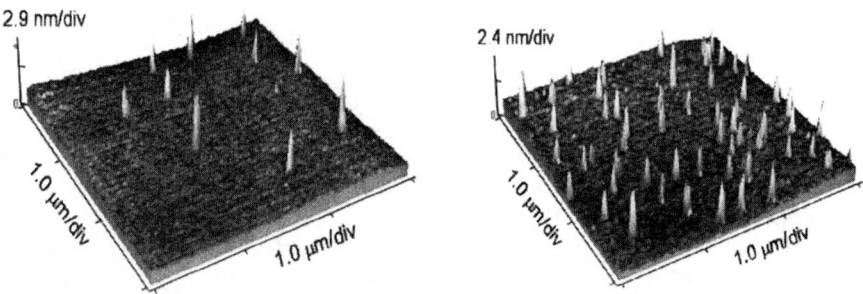

Figure 5. AFM images of oleic acid on a silica surface at a short deposition time (left) and a longer deposition time (right). The number of islands increases with deposition time, but their peak heights are similar in both images. (Reproduced from Reference 21 by permission of the PCCP Owner Societies.)

times. For all deposition times, the mean height of the islands is approximately 28 Å. The observed island heights using AFM of approximately 28 Å are consistent with the measured length of the oleic acid molecule, and the roughness of the silica surface may account for the small difference in values. Therefore, it is likely that the observed islands consist of a single monolayer of oleic acid molecules oriented such that their long axis is perpendicular to the substrate surface.

Oleic acid coatings on polystyrene surfaces

AFM images of the bare polystyrene surface (deposited on silica) and of oleic acid vapor-deposited on the surface are shown in Figure 6. As with the silica substrate, the AFM images show that the oleic acid is forming islands on the surface. Thus, we conclude that oleic acid forms islands when vapor-deposited onto both polar (silica) and nonpolar (polystyrene) surfaces.

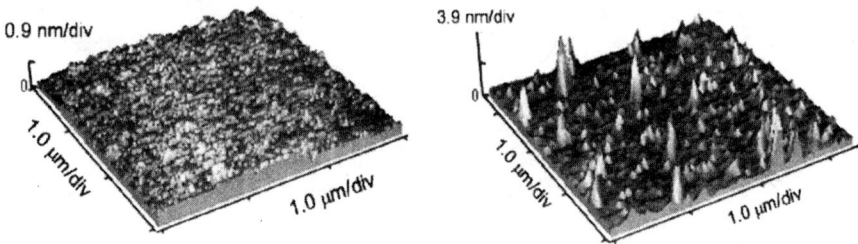

Figure 6. Polystyrene spin-coated on silica (left) and oleic acid deposited on spin-coated polystyrene (right). (Reproduced from Reference 21 by permission of the PCCP Owner Societies.)

Ellipsometry measurements of oleic acid deposition on the polystyrene showed that oleic acid thickness increases with deposition time, similar to the results obtained on silica. The contact angle of water on the bare polystyrene surface is approximately 45°, which is similar to the contact angle for the highest oleic acid coating on the silica surface. As the coating of oleic acid increases on the polystyrene surface, we observe no change in the contact angle. This result demonstrates that the hydrophobicity of polystyrene and of the oleic acid islands is similar.

Discussion

Our discovery that oleic acid forms islands on both silica and polystyrene surfaces resolves both of the unexpected results mentioned in the Kinetics Section. 1) *Constant kinetics over a range of coatings.* As the amount of coating increases, the number of islands increases, but the surface area to volume ratio of the oleic acid coating remains constant. Thus, the kinetics should remain constant. 2) *Decrease in aerodynamic diameter with increased coating.* Formation of islands results in an increase in χ, and so although the average diameter of the particles is increasing with coating, the aerodynamic diameter is not changing significantly.

Our results are relevant to conditions that exist in the atmosphere. It is likely that oleic acid in the atmosphere condenses onto mineral dust or soot, and the oleic acid may arrange itself in islands on these surfaces. It is also possible that surface structure of the oleic acid would be different when it is part of a large mixture of compounds present under typical atmospheric conditions, and so further studies on the morphology of more complex systems would be useful.

Several studies of oleic acid particles with widely varying chemical composition and morphologies have contributed significantly to fundamental understanding of factors that affect aerosol reactivity. It is clear from our results and those of other investigators that different particle morphologies can result in orders of magnitude difference in particle reactivity. Thus, particle morphology, in addition to chemical composition, must be considered when evaluating particle reactivity.

References

1. Worsnop, D. R.; Morris, J. W.; Shi, Q.; Davidovits, P.; Kolb, C. E. A chemical kinetic model for reactive transformations of aerosol particles. *Geophysical Research Letters* **2002**, *29* (20).
2. Morris, J. W.; Davidovits, P.; Jayne, J. T.; Jimenez, J. L.; Shi, Q.; Kolb, C. E.; Worsnop, D. R.; Barney, W. S.; Cass, G. Kinetics of submicron oleic acid aerosols with ozone: A novel aerosol mass spectrometric technique. *Geophysical Research Letters* **2002**, *29* (9).
3. Smith, G. D.; Woods, E.; DeForest, C. L.; Baer, T.; Miller, R. E. Reactive uptake of ozone by oleic acid aerosol particles: Application of single-particle mass spectrometry to heterogeneous reaction kinetics. *Journal of Physical Chemistry A* **2002**, *106* (35), 8085-8095.

4. Nash, D. G.; Tolocka, M. P.; Baer, T. The uptake of O-3 by myristic acid-oleic acid mixed particles: evidence for solid surface layers. *Physical Chemistry Chemical Physics* **2006**, *8* (38), 4468-4475.
5. Knopf, D. A.; Anthony, L. M.; Bertram, A. K. Reactive uptake of O-3 by multicomponent and multiphase mixtures containing oleic acid. *Journal of Physical Chemistry A* **2005**, *109* (25), 5579-5589.
6. Hearn, J. D.; Smith, G. D. Measuring rates of reaction in supercooled organic particles with implications for atmospheric aerosol. *Physical Chemistry Chemical Physics* **2005**, *7* (13), 2549-2551.
7. Katrib, Y.; Biskos, G.; Buseck, P. R.; Davidovits, P.; Jayne, J. T.; Mochida, M.; Wise, M. E.; Worsnop, D. R.; Martin, S. T. Ozonolysis of mixed oleic-acid/stearic-acid particles: Reaction kinetics and chemical morphology. *Journal of Physical Chemistry A* **2005**, *109* (48), 10910-10919.
8. Ziemann, P. J. Aerosol products, mechanisms, and kinetics of heterogeneous reactions of ozone with oleic acid in pure and mixed particles. *Faraday Discussions* **2005**, *130*, 469-490.
9. Hearn, J. D.; Smith, G. A. Ozonolysis of mixed oleic acid/n-docosane particles: The roles of phase, morphology, and metastable states. *Journal of Physical Chemistry A* **2007**, *111* (43), 11059-11065.
10. Knopf, D. A.; Anthony, L. M.; Bertram, A. K. Reactive uptake of O-3 by multicomponent and multiphase mixtures containing oleic acid. *Journal of Physical Chemistry A* **2005**, *109* (25), 5579-5589.
11. Hearn, J. D.; Smith, G. D. Reactions and mass spectra of complex particles using Aerosol CIMS. *International Journal of Mass Spectrometry* **2006**, *258* (1-3), 95-103.
12. Katrib, Y.; Martin, S. T.; Rudich, Y.; Davidovits, P.; Jayne, J. T.; Worsnop, D. R. Density changes of aerosol particles as a result of chemical reaction. *Atmospheric Chemistry and Physics* **2005**, *5*, 275-291.
13. Kwamena, N. O. A.; Abbatt, J. P. Kinetic and product studies of the heterogeneous ozonation reactions of surface-bound polycylic aromatic hydrocarbons. *Abstracts of Papers of the American Chemical Society* **2006**, *231*.
14. Tang, I. N.; Munkelwitz, H. R. Determination of Vapor-Pressure from Droplet Evaporation Kinetics. *Journal of Colloid and Interface Science* **1991**, *141* (1), 109-118.
15. Mykhaylyk, T. A.; Dmitruk, N. L.; Evans, S. D.; Hamley, I. W.; Henderson, J. R. Comparative characterisation by atomic force microscopy and ellipsometry of soft and solid thin films. *Surface and Interface Analysis* **2007**, *39* (7), 575-581.
16. Rader, D. J.; Mcmurry, P. H.; Smith, S. Evaporation Rates of Monodisperse Organic Aerosols in the 0.02-Mu-M-Diameter to 0.2-Mu-M-Diameter Range. *Aerosol Science and Technology* **1987**, *6* (3), 247-260.

17. Hu, X. S.; Shin, K.; Rafailovich, M.; Sokolov, J.; Stein, R.; Chan, Y.; Williams, K.; Wu, W. L.; Kolb, R. Anomalies in the optical index of refraction of spun cast polystyrene thin films. *High Performance Polymers* **2000**, *12* (4), 621-629.
18. Hearn, J. D.; Lovett, A. J.; Smith, G. D. Ozonolysis of oleic acid particles: evidence for a surface reaction and secondary reactions involving Criegee intermediates. *Physical Chemistry Chemical Physics* **2005**, *7* (3), 501-511.
19. Rosen, E. P.; Garland, E. R.; Baer, T. Unpublished Work, **2008**.
20. Hinds, W. C. *Aerosol Technology: Properties, Behavior, and Measurement of Airborne Particles;* 2nd ed.; Wiley-Interscience: New York, 1999.
21. Garland, E. R.; Rosen, E. P.; Baer, T. *Physcial Chemistry Chemical Physics,* **2008**, in press.
22. Kubono, A.; Yuasa, N.; Shao, H. L.; Umemoto, S.; Okui, N. Adsorption characteristics of organic long chain molecules during physical vapor deposition. *Applied Surface Science* **2002**, *193* (1-4), 195-203.

Chapter 3

The Characteristics and the Cytotoxic Effects of Particulate Matter in the Ambient Air of the Chiang Mai-Lamphun Basin in Northern Thailand

Narongpan Chunram[1], Usanaee Vinitketkumnuen[2], Richard L. Deming[3], and Richard M. Kamens[4]

[1]Department of Applied Science, Faculty of Science and Technology, Chiang Mai Rajabhat University, Chiang Mai 50300, Thailand
[2]Department of Biochemistry, Faculty of Medicine, Chiang Mai University, Chiang Mai 50200, Thailand
[3]Department of Chemistry and Biochemistry, California State University at Fullerton, Fullerton, CA 92834
[4]Department of Environmental Science and Engineering, University of North Carolina, Chapel Hill, NC 27599

There are growing concerns about the deteriorating air quality in the Chiang Mai-Lamphun Basin in Northern Thailand resulting from accelerating development and construction, increasing industrial emissions, expanding transportation, extensive open burning and the continuing drought. Recent studies of the levels of fine particulate matter ($PM_{2.5}$ and PM_{10}) in the Basin have highlighted their seasonal variations, correlations with other pollutants, and toxicological properties. These are reviewed and implications for future research are summarized.

Introduction

The effects of airborne pollutants on human health have been widely recognized and extensively examined. Results of numerous studies around the world have documented the increases in respiratory distress, pulmonary disease and cancer, leading to increased hospital admissions and higher mortality associated with a variety of pollutants (*1-10*). Of special concern is particulate matter (PM) of less than 10 microns (PM_{10}), and, especially, of the fine particles of less than 2.5 microns ($PM_{2.5}$) (*11*). Air pollution problems in urban areas are especially significant because of the large numbers of people that can be affected by the numerous and varied sources of pollutants in these densely settled areas. In addition, rural areas surrounding cities often have significant impact on urban centers because of the transport of materials under varying climatological, atmospheric and geophysical conditions.

This review highlights some of the important data and conclusions of recent investigations (*12-15*) that focus on the rapidly growing Chiang Mai-Lamphun basin in Northern Thailand (*Figure 1*). The work reviews the data on the levels and distribution of <10 μm and <2.5 μm particulate matter, PM_{10} and $PM_{2.5}$, respectively, along with the mutagenicity and cytotoxicity of samples collected on filters in the Chiang Mai-Lamphun Basin, providing information about the extent of the problem and the potential health effects.

The adjacent provinces of Chiang Mai (population 1.5 million) and Lamphun (0.45 million) are situated in the northern part of Thailand in the mountainous and forest-covered region near Myanmar (Burma) and Laos. The abundant rainfall and fertile hills and valleys in the region support agricultural crops such as rice, sugar cane, citrus, coffee and vegetables. It is one of the most attractive year-round tourists destinations in Southeast Asia. The Chiang Mai provincial capital city (metropolitan population 0.7 million) and nearby provincial capital city of Lamphun incorporate urban, suburban and industrial areas and are located in a basin between mountain ranges. This fast-growing region that is experiencing many environmental stresses.

The three distinct climatic seasons in Northern Thailand have significant effects on air quality. The rainy season extends from 1 June to 15 October, the winter, or dry, season is from 16 October to 15 March, and the summer season is from 16 March to 31 May. The basin geography creates the conditions for frequent thermal inversions at certan times of the year, especially in the cooler winter months, trapping pollutants and limiting air flow and mixing. The year-round heavy vehicular traffic (not subject to inspection and much of it relying on diesel fuel) and increasing industrial development make noticeable contributions to the atmosphere by lowering the air quality. Cycles of construction of roads, superhighways and buildings make periodic and significant contributions to dust and particulate levels.

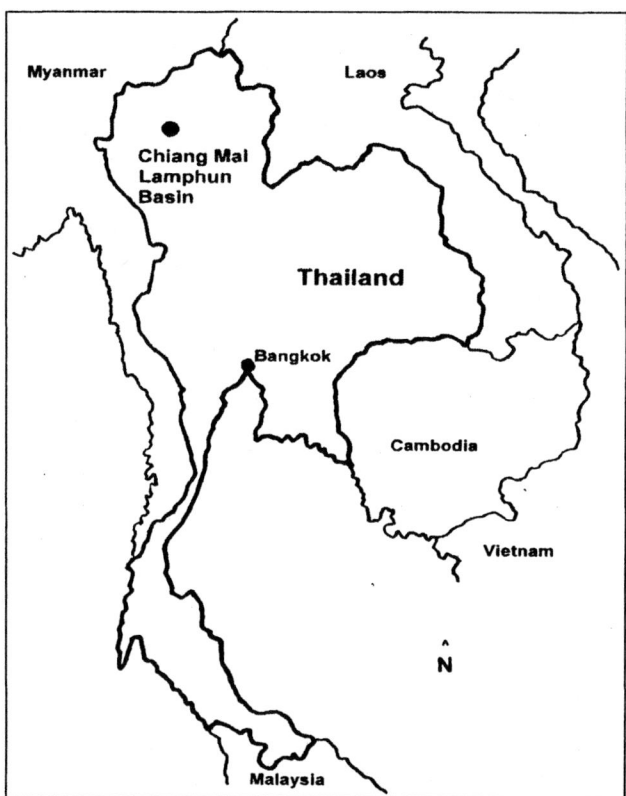

Figure 1. Map of Thailand highlighting the Chiang Mai-Lamphun Basin.

It is common agricultural practice in the months of January through April to burn dry crops to reduce the vegetation and prepare the fields for planting. Also, frequent forest fires in the dry (winter) season are often set by villagers in order to clear dense underbrush in forested areas of the surrounding mountains. Contributions from fires in neighboring Myanmar, and even from as far-away as Indonesia, add to the particulate load in the atmosphere. Although air quality is generally best during the rainy season, there may still be periods of locally high levels of pollutants. As a result of all of these conditions, there are frequent increases in hospital admissions and significant health threats from the high levels of pollutants many days of the year.

A recent study (*16*) found that lung cancer rates in Chiang Mai, Thailand, were found to be the second highest in the world. The high incidence of cancer, especially of lung cancer, in Chiang Mai relative to other places in Thailand was reported in 1993 (*17*) and considerable effort has been devoted to examining the contributions from components in polluted air. In 2002, a study (*12*) was published on the levels and mutagenecity of airborne PM_{10} and $PM_{2.5}$, in urban areas of Chiang Mai using data collected during the period March 1998 to October 1999.

One of the greatest challenges in determining the specific agents that cause the detrimental effects of air pollution on organisms is the complexity of the samples. It is clear that some of the most potent toxins are the polycyclic aromatic hydrocarbons, PAH's, but a number of other organic and inorganic compounds are also capable of inducing cellular changes. Initial screening studies usually are usually based on genotoxic or mutagenic properties of a mixture of materials extracted from filter samples by applying extracts to specific cells or cell lines (*18-28*). Studies focusing on the levels and impact of particulate matter and other pollutants in the Chiang Mai – Lamphun Basin are summarized below, along with their key findings.

Experimental

Particulate samples were collected for 24-hour periods using mini-volume air samplers (AIRmetrics MiniVol portable samplers, USA) on 47 mm fiber-film filters (Pallflex, USA) with <2.5 μm and <10 μm cutoffs for $PM_{2.5}$ and PM_{10} measurements, respectively, operating at 2.5 L/min.. Sampling sites in Chiang Mai included a residential apartment and a research laboratory, both near heavy traffic areas a rural area with limited direct automotive traffic exposure, a commercial market on a busy street, a site in an area known to have a high incidence of lung cancer. In Lamphun, they included both industrialized and urban areas. For daily measurements, filters were conditioned in an electronic dessicator at 25°C and relative humidity of 50% before and after sample collection and weighed on a micrometric balance (Sartorius AG, Germany) to the nearest 0.001 μg. Meterological data (wind direction, wind speed and humidity levels) were collected with standard equipment mounted near selected sites, and levels of SO_2, NO_2 and O_3 used for correlations were obtained from local and provincial governmental pollution monitoring agencies (*13*).

For toxicity studies following mass measurements (*14*), filters collected over one month's time were were cut into small pieces, pooled, and sonicated for 15 minutes in 200 mL of dichloromethane. After addition of anhydrous sodium sulfate, the extract was filtered through Whatman No. 41 filter paper. Sonication was repeated two more times and the samples were dried at 35°C on a rotary evaporator. For mutagenicity studies on the residue, the Ames test utilized *Salmonella typhimurium* strains TA98 and TA100, with proper controls, as described previously (*12*). Revertant colonies per plate were counted and the toxic effects were examined under a stereomicroscope.

For cell viability and DNA fragmentation studies (*15*), the residue was quantitatively resolved in culture media and filtered through a Millipore membrane to ensure sterility and stored in the dark at 4°C in sealed vials. Human alveolar type II-derived cell line A549 and alveolar macrophages MH-S were grown on F-12KDMEM(1:1) and RPMI-1640 media, respectively, supplemented with 10% fetal bovine serium (FBS), 100 U/mL penicillin, and 10 mg/mL streptomycin in a humidified atmosphere at 37°C and 5% CO_2. The 3-(4,5-dimethylthiazol-2-yl) 2,5-diphenyltetrazolium bromide (MTT) assay was used for cell viability measurements, as described earlier (*12*). DNA fragementation was examined by agarose gell electrophoresis with ethidium bromide staining, following lysis in buffers, incubation with RNase A and then Proteinase K, precipitation with NaCl, centrifugation, washing, and dissolution in TE buffer.

Results and Discussion

The daily levels of $PM_{2.5}$ and PM_{10} were measured at several sites in the Basin covering the period of June 2004 through May 2006. The results from the residential sampling site are illustrated in Figure 2 for the period of June 2004 to May 2005, a typical annual pattern for all sites.

Average seasonal variations at two sampling sites are summarized in Table I for the period June 2004 to May 2006. As noted in the table, the end of the winter (dry) season and the beginning of the summer season show extremely high levels of particulate matter. The winter season is characterized by cold weather and frequent thermal inversions with limited dispersion from low wind speeds, and increased biogenic emissions from local burning. During those times, average $PM_{2.5}$ mass concentrations at various sites ranged from 29.9 (rural area) to 44.5 µg/m³ (commercial market) at a miniumum, and the maximum 24-hour levels ranged from 115.1 µg/m³ (research laboratory) to 257.5 µg/m³ (high lung cancer area). The new 2006 USEPA 24-hour standard for $PM_{2.5}$ of 35 µg/m³ was exceeded 30-50% of the days during the study periods.

A number of studies have focused on meteorological contributions to particulate matter and correlations with other constituents (*29-31*). Correlations of $PM_{2.5}$ with PM_{10} with other atmospheric constituents, SO_2 and NO_2, in the Chiang Mai-Lamphun Basin are positive and high, with significant variation for different wind directions and speeds. With respect to ozone, O_3, the unusually high level in summer is consistent with the suggestion (*32*) that atmospheric chemistry, rather than transport, may be responsible for these high summer concentrations. High concentrations of particulate matter occur with moderate southerly winds (6-12 m/s blowing from the industrial areas and the international airport, moderate temperature (20-25 °C), and low humidity (near 50%) during winter.

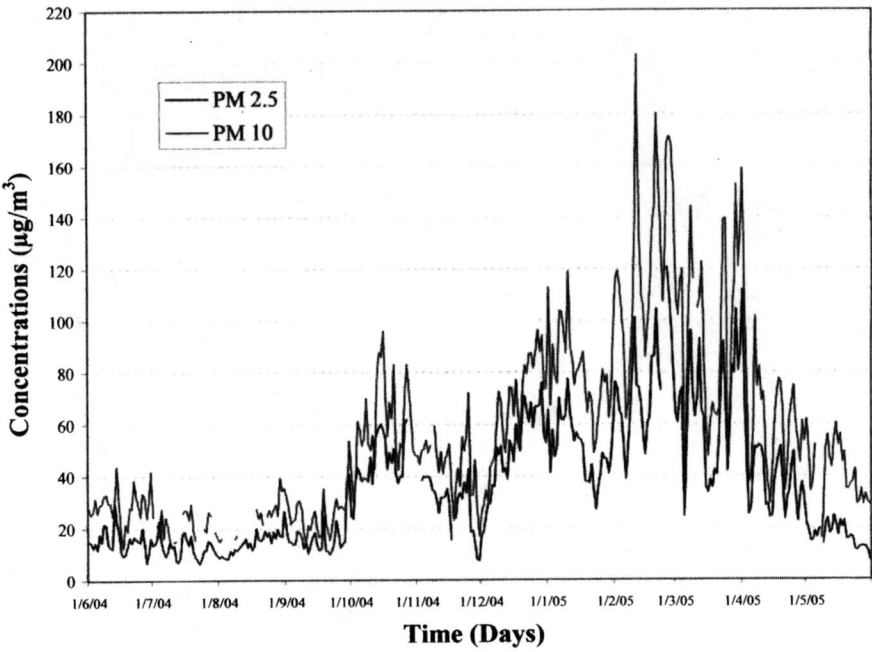

Figure 2. Daily levels of $PM_{2.5}$ and PM_{10} ($\mu g/m^3$) at a residential site during the period of June 2004 (1/6/04) through May 2005 (31/5/05). (Modified from Reference 13.)

The 2002 study (*12*) demonstrated mutagenicity to *Salmonella typhimurium* strain TA100 without activation for extracts of airborne particulate matter collected in winter in Chiang Mai. Sites included a fifth-floor balcony of a multidisciplinary building Chiang Mai University, the ground level outdoor patio of a residential home near the same high-traffic area, and a downtown market area. Monthly average $PM_{2.5}$ levels ranged from 17 to 138 $\mu g/m^3$ and PM_{10} levels ranged from 27 to 173 $\mu g/m^3$, and daily levels in February and March reaching as high as 208 $\mu g/m^3$. Mutagenicity increased in the presence of metabiolic activation (S9 mix) and appeared to track particle concentrations. Indirect mutagenicity in airborne particulate matter extracts were detectable during the winter months, October to March. Direct-acting mutagenicity was detected at one site, and mutagenic activity was higher in the presence of enzyme activation.

The more recent study (*14*) utilizing *Salmonella typhimurium* provides additional information about the toxicity of particulate matter in the 24-hour filter samples collected in urban areas of Chiang Mai from June 2004 to May 2006. With both indoor and outdoor sampling, the types of mutations induced by the complex environmental samples were found to be the same types as those

Table I. Seasonal averages of $PM_{2.5}$ and PM_{10} for two sites in the Chiang Mai-Lamphun Basin from June 2004 to May 2006

Season	$PM_{2.5}$	PM_{10}	SO_2	NO_2	O_3
Rainy					
Ave	17.8	31.2	1.9	10.7	16.7
Max	60.2	92.5			
Min	6.4	12.6			
Std Dev	10.6	20.6			
Winter					
Ave	52.3	96.1	3.3	31.2	40.4
Max	120.5	202.9			
Min	7.5	18.5			
Std Dev	20.6	47.4			
Summer					
Ave	36.4	60.9	2.8	19.7	57.6
Max	111.6	158.7			
Min	7.7	16.2			
Std Dev	23.6	31.5			

NOTE: All numeric values are in $\mu g/m^3$
SOURCE: Modified from Reference 13

in humans exposed to the same environmental mutagens. Indirect-acting mutagenicities were detected in both sites in all samples. The seasonal trends were higher levels in the winter months and lower to undetectable levels in the summer and rainy seasons. It has been noted that organic extractable matter from air particles and different combustion sources are carcinogenic in animals and mutagenic in short-term bioassay tests (23). Also, vehicle emissions have been shown account for most of the mutagenic activity associated with air particles in urban areas (24). Additional contributions from cooking and tobacco smoke are likely. Finally, during high ozone (O_3) and NO_2 episodes, OH radicals can facilitate formation of nitro-PAHs which have higher mutagenicity than PAHs (25).

Another study (15) using human cell line A549 (alveolar type II-derived cell line) and MH-S alveolar macrophages, involved screening for cytotoxicity and apoptosis (cell death) induction in cultured lung cells and macrophages, using 24-hour filter samples for the period from July 2004 to May 2005. MTT (3-(4,5-dimethylthiazol-2-yl) 2,5-diphenyltetrazolium bromide) assays in human lung cells and alveolar macrophages were carried out, along with DNA fragmentation to detect apoptosis induction. The $PM_{2.5}$ and PM_{10} components were cytotoxic to lung cells and alveolar macrophages, and, at one site in Lamphun province, apoptosis of alveolar macrophages was induced. High cytotoxicity (about 40%) was observed for 72-hr exporsure to $PM_{2.5}$ and PM_{10} extracts, with larger effects on MH-S than on A549 lung cells. Apoptosis of alveolar macrophages was observed at one site in Lamphun province for samples collected in May, near the end of the summer season. These results are important because they imply that exposure to ambient particulate matter may be related to short-term and as well as long-term respiratory and health problems.

Implications

The studies in the Chiang Mai-Lamphun basin described above suggest that there is a growing need for regular and continuous monitoring of $PM_{2.5}$ and PM_{10} levels. Particulate matter formation is not limited to only urban areas, as can be clearly seen by high levels of smoke and haze in much of the region due to open burning and dry conditions. Sampling sites outside of the central city have noticeably high levels of particulates, as well. If pollution regulations are to be established and enforced, and if the health of the people is to be improved, monitoring to obtain solid ongoing scientific information must continue and results must be made available to government officials and local planners.

Additional toxicological research is needed to more closely examine detrimental effects of the $PM_{2.5}$ and PM_{10} on human health. More detailed analytical measurements involving identification and quantitation of particular chemical species need to be developed in order to identify specific compounds in the local particulate matter that are responsible for the adverse health effects and facilitate the fingerprinting of possible sources. Finally, improved regulation and monitoring, greater public awareness and education, and expanded involvement of public health professionals and local leaders will contribute greatly to solving this ongoing problem and lead to improved health.

References

1. Dockery, D. W.; Pope, C. A.; Xu, S.; Spengler, J. D.; Ware, J. H.; Fay, M. E.; Ferris, B. G., Jr.; Speizer, F. E. *New England J. Med.* **1993**, *329*, 1753-1759.
2. Harrison, R. M.; Yin, J. *Sci. Total Environ.* **2000**, *249*, 85-101.
3. Becker, S.; Soukup, J.M.; Gallagher, J.E. *Toxicol. In Vitro.* **2002**, *16*, 209-210.
4. Schwatrz, J. *Epidemiology,* **1996**, *7*, 20-28.
5. Schwartz, J.; Dockery, D.W.; Neas, L. M. *J. Air Waste Manage. Assoc.* **1996**, *46*, 927-939
6. Trakultivakorn, M. *Asian Pacific J. Allergy Immunol.* **1999**, *17(4)*, 243-248.
7. Pope, C.A.; Burnett, R.; Thun, M.J.; Calle, E.E.; Krewskik, D.; Ito, K.; Thurston, G.D. *J. Am. Med. Assoc.* **2002**, *287*, 1132-1141.

8. Samat, J. M.; Dominici, F.; Curriero, F. C.; Coursak, I.; Zeger, S. L. *New Engl. J. Med*, **2000**, *343*, 1742-1749.
9. Ostro, B.; Chestnut, L.; Vichit-vadakan, N.; Laixthai, A. *J. Air Waste Manag. Assoc.* **1999**, *49*, 100-107.
10. Neas, L.M. *Fuel Process. Tech.* **2000**, *65-66*, 55-67.
11. World Health Organization (WHO), Report on a WHO Working Group, Bonn, Germany, 13-14 January 2003, Regional Office for Europe, Copenhagen.
12. Vinitketkumnuen, U.; Kalayanamitra, K.; Chewonarin, T; Kamens, R. *Mutat. Res.*, **2002**, *519*, 121-131.
13. Chunram, N.; Vinitketkumnuen, U.; Deming, R.; Kamens, R *J. Yala Rajabhat Univ.*, **2007**, *2(1)*, 1-13
14. Chunram, N.; Kamens, R.; Deming, R., Vinitketkumnuen, U. Mutagenicity of outdoor and indoor PM 2.5 from urban areas of Chiang Mai, Thailand. *Chiang Mai Med. J.* **2007**, *46(1)*, 1-11.
15. Vinitketkumnuen, U.; Taneyhill, K. P.; Chewonarin, T.; Chunram, N.; Vinitketkumnuen, A.; Tansuwanwong, S. *CMU J. Nat. Sci.* **2007**, *61(1)*, 1-10.
16. Heepchantree, W.; Paratasilpin, T.; Kangwanpong, D. *Mutat. Res.*, **2005**, *587*, 134-139.
17. Vatanasapt, V.; Martin, N.; Sriplung, H.; Chindavijak, K.; Sontipong, S.; Sriamporn, S., Parking, D. M.; Ferley, J. Cancer in Thailand 1988-1991, IARC Technical Report; No. 16: Lyon.
18. de Martinis, B. S.; Kado, N.Y.; de Carvalho, L. R.; Okamoto, R. A.; Gundel, L. A. *Mutat. Res.* **1999**, *446*, 83-94.
19. Daya, U.; Vijayalakshmi, P.; Andrew, G.; David, W. *Am. J. Respir. Cell. Mol. Biol.* **2003**, *29*, 180-187.
20. Kado, N. Y.; Langley, D.; Eisenstadt, E. *Mutat. Res.* **1983**, *121*, 25-32.
21. Cassoni, F.; Bocchi, C.; Martinio, A.; Pinto, G.; Fontana, F.; Buschini, A. *Sci. Total Environ.* **2003**, *324*, 79-90.
22. du Four, V. A.; van Larabeke, N.; Janssen, C. R. Mutat. Res. **2003**, *525*, 43-59.
23. Grimmer, G.; et. al., *Cancer Lett.* **1987**, *37*, 173-180.
24. Lewis, C.; Baumgardner, R.; Claxton, L.; Lewtas, J.; Stevens, R. *Environ. Sci. Technol.* **1988**, *22*, 968-971.
25. Finlayson-Pitts, B. J.; Pitts, J. N. Biological properties of PAHs and PACs, in *Chemistry of the Upper and Lower Atmospheres*, Academic Press, San Diego, 2000; pp. 466-474.
26. Ruchirawat, M.; Mahidol, C., Tangjarukij, C. *Sci. Total Environ.* **2002**, *287*, 121-132.
27. Matsushita, H.; Tabucanon, M. S.; Koottatep, S. Proceedings of the third joint conference of air pollution studies in Asian areas, **1987**, pp. 304-285.

28. Ruchirawat, M.; Navasumrit, P.; Settachan, D. *Ann. N. Y. Acad. Sci.* **2006**, *1076*, 678-690.
29. Bari, A.; Ferraro, V.; Wilson, L.R.; Luttinger D.; Husain, L *Atmos. Environ.* **2003**, *37*, 2825-2835.
30. Hien, P. D.; Bac, V. T.; Tham, H. C.; Nhan, D. D. Vinh, L. D. *Atmos. Environ.* **2002**, *36*, 3473-3484.
31. DeGaetano, A.; Doherty, O. *Atmos. Environ.* **2004**, *38*, 1547-1558.
32. Cox, W. M.; Chu, S-H. *Atmos. Environ.* **1996**, *30*, 2615-2625.

Chapter 4

Toluene Decomposition on Water Droplets in Corona Discharge

Ying Kang and Zucheng Wu*

Department of Environmental Science and Engineering, Zhejiang University, Hangzhou 310027, Peoples Republic of China
*Corresponding author: wuzc@zju.edu.cn; sunway9928@163.com

Heterogeneous reactions on multiphase media have the potential to play a major role in the decomposition of volatile organic compounds. Making a distinction between reactions in the gas phase and those occurring in liquid droplets is convenient for understanding the degradation mechanism of toluene which has served as a model pollutant. In a corona radical shower reactor, strong oxidants such as hydroxyl radicals were efficiently produced to degrade toluene in the gas phase. H_2O_2 and O_3 appeared in both phases due to the recombination of hydroxyl radicals; some organic intermediate decomposition products of toluene were found in the liquid phase. The metal ions present in the nebulized solution probably as catalysts to facilitate the decomposition of organic intermediates when they were introduced into a corona zone in the form of a mist. These results gave us evidence that toluene degradation first occurred in the gas phase. The oxidants and soluble organic intermediate products probably dissolved into the atomized droplets containing metal ions and underwent further degradation. It also suggests that the mist played an important role in the homogenization of the heterogeneous catalysis of the multiphase reactions. The catalytic ability of both Co^{2+} and Mn^{2+} in toluene degradation was similar to that of Fe^{2+}.

© 2009 American Chemical Society

The emission of volatile organic compounds (VOCs) into the atmosphere from various industrial processes is an increasing environmental and social concern in recent years (*1*). Some of them are carcinogens and others cause respiratory disorders. Therefore, effective processes must be developed facilitate degradation of VOCs.

Non-thermal plasma (NTP) technology has proven to be effective in degrading aromatic compounds and other VOCs (*2-4*). The non-equilibrium conditions of an NTP are manifested in the fact that the electrons are selectively accelerated by a strong external electric field resulting in electron temperatures above 10^4 K while the temperature of the neutral gas molecules remains practically "cold". The high energy electrons excite the radical source molecules such as H_2O and O_2, producing active radicals like ·OH. Generally speaking, strong oxidative radicals, such as ·OH, have played important role in the oxidation of pollutants (*5*). The ensuing chain reactions afterward between the radicals and the VOC molecules will propagate to form stable, less toxic compounds. The ideal products of decomposition of toxic compounds are CO_2 and H_2O.

Corona radical shower is one kind of NTP discharge that is capable of effective production of radicals between a nozzle electrode and a ground electrode under excitation in a high voltage DC field. Improving the generation rate of active radicals is a promising avenue of exploration to enhance pollutant degradation rates. Although the most efficient way to achieve this aim is increasing the applied voltage between the two electrodes, a spark will occur when the breakdown voltage is reached. Hence attempts to combine elements of two existing technologies, NTP and catalysis have been carried out by many researchers (*6, 7*). Existing combinations of the two technologies usually employ heterogeneous catalysis with the catalyst loaded on a granular support which completely fills the reactor to make an in-plasma catalysis reactor or a packed-bed reactor (catalyst without plasma) for the post-treatment of the plasma effluent (post-plasma catalysis reactor). These methods, however, have substantially increased gas resistance and a corresponding reduction in the effective volume of the reactor. And these reactors can not be used in a corona discharge because of the difficulty of forming a smooth and steady corona in the reactor. Therefore, in the corona radical shower reactor, the use of a packing catalyst may not be so suitable and perhaps the introduction of liquid phase catalysts into the corona radical shower could overcome the limitations identified above.

A literature search showed few references to gas phase reactions catalyzed by liquid phase catalysts owing to the difficulty of obtaining a homogeneous distribution in the reaction chamber (*8*). The degradation of gas phase pollutants would be efficiently enhanced if liquid phase catalysts could be introduced into the gas phase reaction through a suitable distribution mechanism and subsequently separated from the reaction products.

The Fenton reagent is potentially an effective system since it is an environmentally friendly homogeneous liquid catalyst. Within it, strongly oxidative ·OH is formed from H_2O_2 by the catalysis of the bivalent cation in acid solution (9, 10). The chain reactions between the radicals produced and target molecules will be maintained until stable compounds are formed. If the Fenton reagent were to be introduced into the gas phase, its advantages of catalysis oxidation might be exploited (11).

In the work described here, the feasibility of homogenization of heterogeneous catalysis is explored by means of aerosol dispersion. The bivalent cation solution, used as a catalyst, was combined with an electrode gas, O_2, to form a mist in a nebulizer and then introduced into the reaction zone. Toluene was used as an example of many long-lived environmental and potentially carcinogenic pollutant. The separation and further recycling of the catalyst after the reaction was performed with a catalyst deentrainment column.

Toluene decomposition in gas phase and water droplets

Reactions in gas phase

Clearly the electrode gas O_2 can be activated, and active species like ·OH, ·O can be formed by electron excitation which dependeds on the magnitude of the supply voltage (12).

$$O_2 + H_2O + e^* \rightarrow ·O + ·OH + HO_2· + H_2O_2 \qquad (1)$$

$$O_2 + O_2 + e^* \rightarrow ·O + O_3 \qquad (2)$$

The hydroxyl radicals in eq(1) was clearly identified in our previous works through electron spine resonance (13). The quantity of oxidative species is seen to increase with an increase in applied voltage. Figure 1 (a) shows the O_3's concentration without addition of toluene to the reactor, and Figure 1 (b) presencts the results of the process that includes toluene in the voltage range of 4 kV to 22 kV. It can be seen from the Figure, there were nearly no O_3 detected at voltage below 10 kV, a remarkable increased amount of O_3 was produced with the increasing of the applied voltage, especially at the voltage higher than 12 kV whether the toluene was added or not. Prolonging the residence time is propitious to the production of O_3.

There are four peaks in the near ultraviolet region that can be distinguished (Figure 2), at wavelengths of 315.3 nm, 335.2 nm, 356.1 nm and 379.4 nm respectively. Those UV emissions result from the excitation of nitrogen atoms (14). The production of •OH can be enhanced through the reactions of singlet oxygen atom from the photolysis of O_3 (15):

$$O_3 + hv \rightarrow O(^1D) + O_2 \qquad (3)$$

$$O(^1D) + H_2O \rightarrow \cdot OH + \cdot OH \qquad (4)$$

$$\text{Net: } O_3 + H_2O \rightarrow 2 \cdot OH + O_2 \qquad (5)$$

The oxidative species in the gas phase reached a maximum concentration of 900 mg m^{-3} at a voltage of 29 kV. However, when 1000 mg m^{-3} toluene was added into the gas flow, the concentration of oxidative species fell from 900 mg m^{-3} to 750 mg m^{-3} at the same voltage. This shows that some of the oxidant generated by the plasma activation of the electrode gas was consumed in the degradation of toluene. The following reaction may occur in the gas phase toluene decomposition.

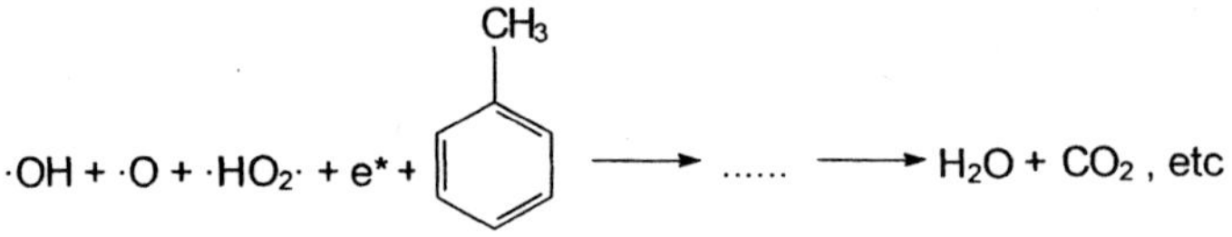

$$\cdot OH + \cdot O + \cdot HO_2 \cdot + e^* + C_6H_5CH_3 \rightarrow \cdots \rightarrow H_2O + CO_2, \text{ etc} \qquad (6)$$

Reactions in droplets

Analysis of samples from droplets showed over a dozen degradation products present, and these were in identified from the HPLC chromatograms (Figure 3). However, toluene is less soluble in water it has only been detected in the gas phase, not the liquid phase. This observation indicates that toluene must first be oxidized in the gas phase to form intermediate products due to the oxidization by active species such as •OH. These hydrophilic substances easily dissolve into the atomized droplets and then are further oxidized owing to the co-dissolved H_2O_2. Reducing the diameter of the droplets will produce a larger surface area and therefore will be beneficial to the diffusion capability.

Soluble organic substances such as benzoic acid, phenyl aldehyde, phenylcarbinol (benzyl alcohol), para-hydroxybenzoic acid, phenol, fumaric acid and oxalic acid were detected in the liquid sample (see Figure 3) using HPLC. The ascending order of bond strengths in the toluene molecule, are: C-H bond on the methyl, C-C bond between methyl and benzene ring, C-H bond on the benzene ring and C-C on the benzene ring. This means the oxidation reaction on the methyl is much more likely to occur than that on the ring itself. Clearly, phenyl aldehyde, phenylcarbinol and benzoic acid are directly formed from toluene by hydroxyl radical adducts.

$$Ph\text{-}CH_3 + \cdot OH \rightleftharpoons Ph\text{-}CH_2OH + \cdot H \quad (7)$$

$$Ph\text{-}CH_2OH + \cdot OH \rightleftharpoons Ph\text{-}CHO + \cdot H \quad (8)$$

$$Ph\text{-}CHO + \cdot OH \rightleftharpoons Ph\text{-}COOH + \cdot H \quad (9)$$

In the chromatogram, both phenylcarbinol and benzoic acid are present in small amounts, indicating that phenylcarbinol is easily oxidized to phenyl aldehyde. The benzoic acid can be further degraded by addition of •OH to form hydroxybenzoic acid and lead to small amounts of benzoic acid appearing. The ring breaking reaction will take place owing to the further oxidation of hydroxybenzoic acid resulting in the formation of acid fumaric and oxalic acid which are clearly shown in the HPLC chromatogram. This indicates that strong oxidants like •OH and HO_2• will first oxidize toluene to phenylcarbinol, benzoic acid and para-hydroxybenzoic acid, then fumaric acid and oxalic acid and, finally, carbon dioxide.

Figure 4 shows the possible principal mechanism of the toluene degradation. In the gas phase of the corona radical shower, a portion of the toluene (represented by RH) is firstly attacked by the strong oxidative radicals •OH and some intermediate products are produced (represented by HORH•). These adducts of toluene, much more soluble, will be further broken down to their final products (CO_2 and H_2O).

Even though the reactions taking place in the corona radical shower may be very complicated, Eqs. (1) and (2) would be the main reactions that are expected in the gas phase discharge in the presence of O_2 and H_2O. Due to the short lifetime of radicals and analytical difficulties, only the total oxidant yield can be

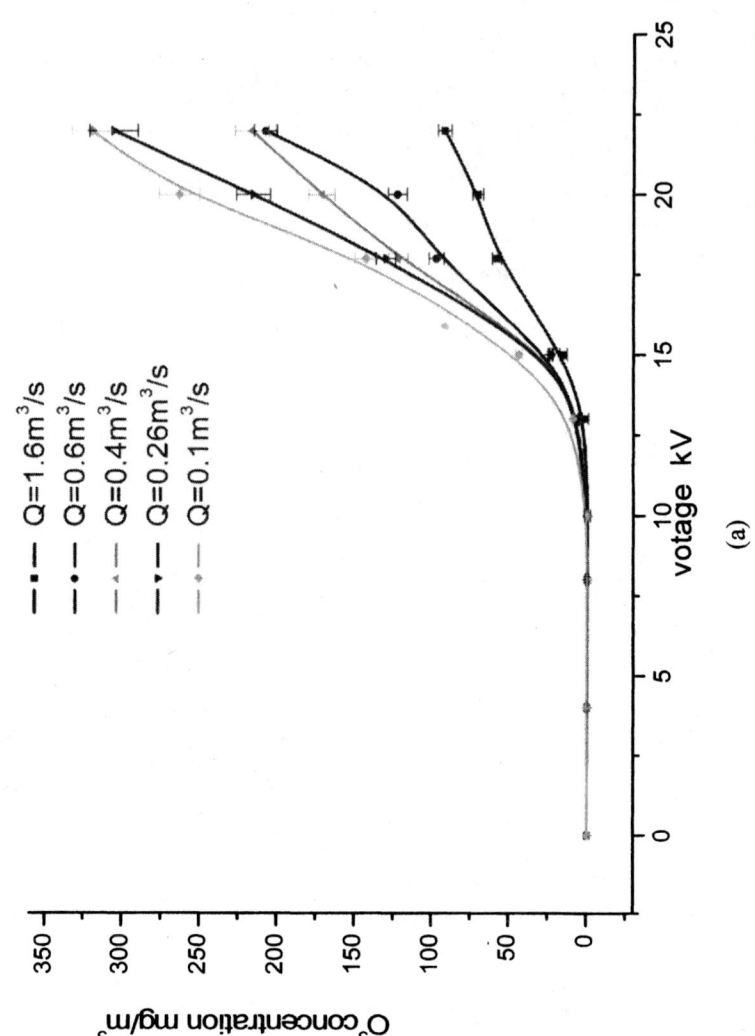

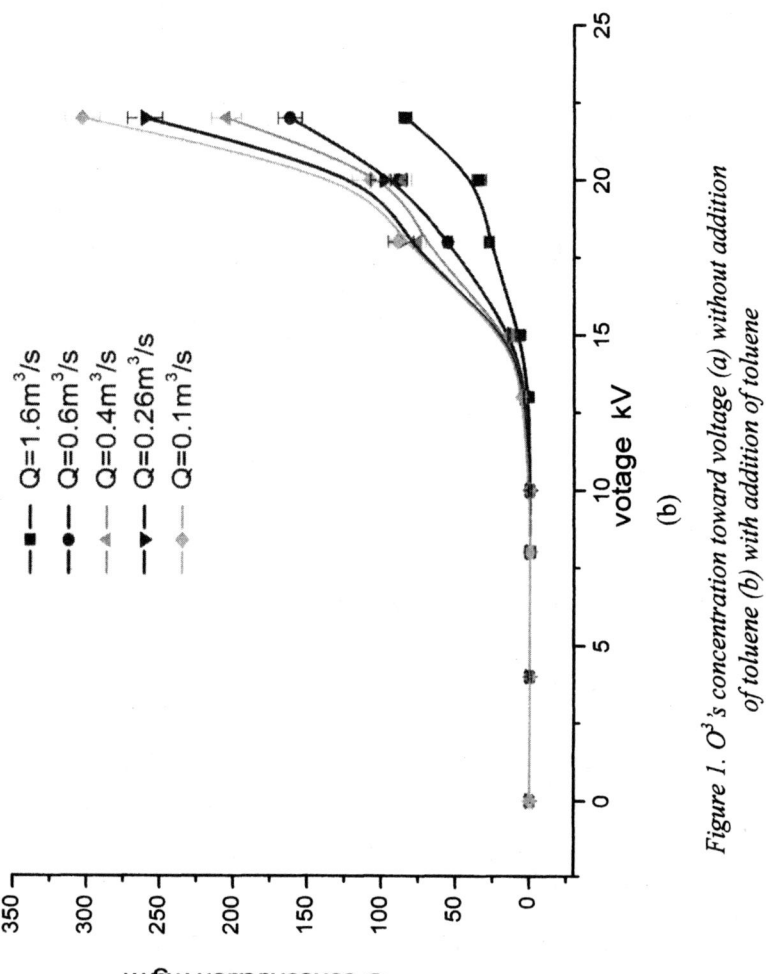

Figure 1. O_3's concentration toward voltage (a) without addition of toluene (b) with addition of toluene

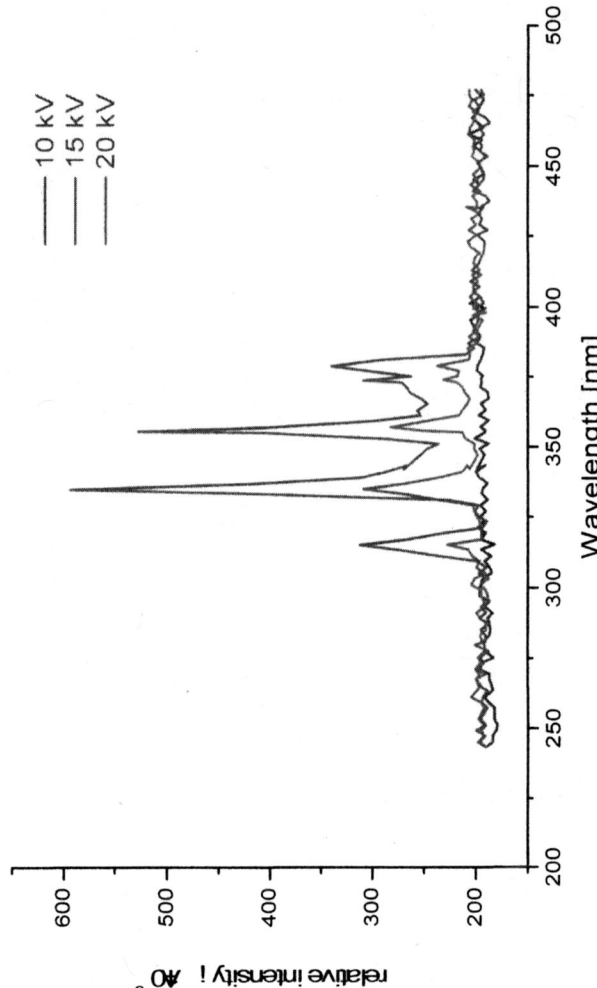

Figure 2. Spectrum of non-thermal discharge reactor at different voltage

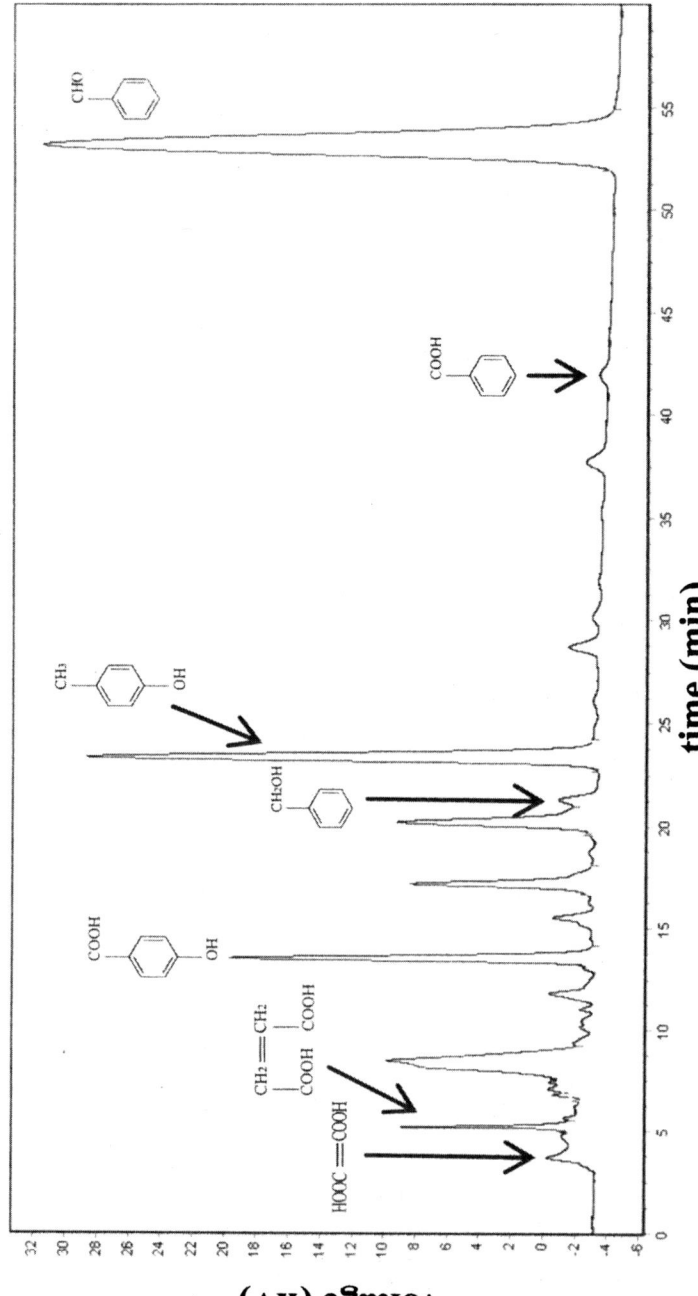

Figure 3. HPLC analysis of intermediate product of toluene decomposition in liquid sample (collected from catalyst retrieval chamber)

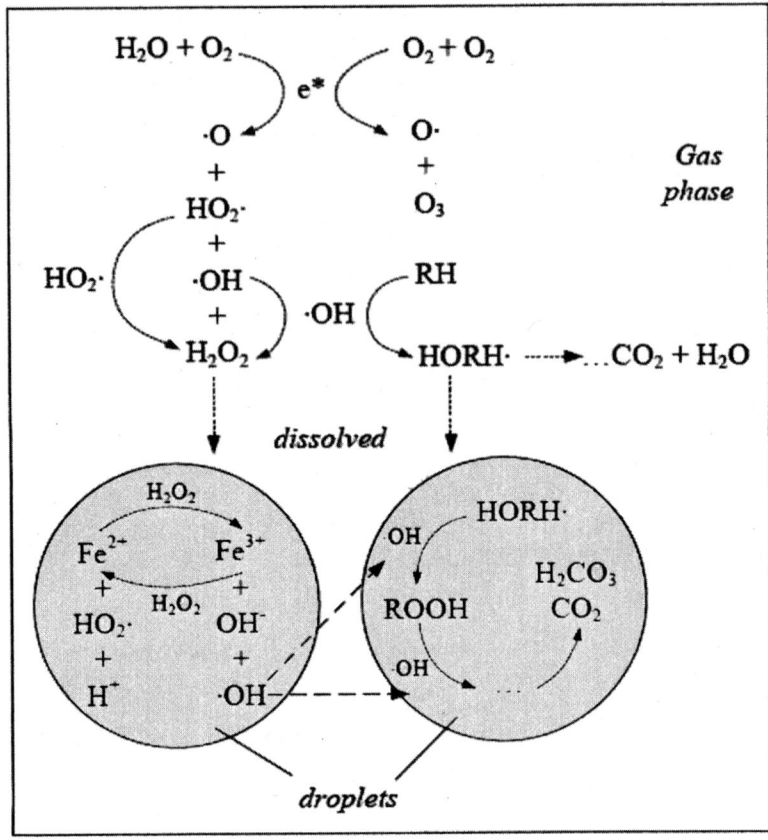

Figure 4. Toluene degradation route by Fe^{2+} catalyst in corona radical shower

quantified. From Figure 2, it is clearly implied that the strong radical •OH has an affect on the toluene degradation.

A group of side reactions, detailed in Eq. (10) and (11), also occur result in chain termination. More stable and less oxidative H_2O_2 was formed in the gas phase with increased consumption of active radicals.

$$HO_2\bullet + HO_2\bullet \rightarrow H_2O_2 + O_2 \qquad (10)$$

$$\bullet OH + \bullet OH \rightarrow H_2O_2 \qquad (11)$$

Homogeneous enhancement of toluene decomposition with Fe^{2+}

Figure 5 shows the result of the toluene decomposition with respect to Fe^{2+} catalysis in the plasma under different voltages. The total oxidant, including H_2O_2, O_3 and gas phase oxidative radicals, was determined. The maximum decomposition of toluene is dependant on the applied voltage. Sixty five percent (65%) of the toluene was removed at the voltage of 29 kV, and this increased to 92% removal efficiency when catalyst was added.

After 0.05 mmol L^{-1} Fe^{2+} solution was introduced into the plasma reactor with the electrode gas, a further decline of the oxidants' concentration was observed. A more rapid removal of toluene was observed than in the case where no catalyst was present. It is clear that the Fe^{2+} acted as a catalyst during the toluene degradation process. This is probably because H_2O_2 passed from the gas phase into the liquid phase through the gas-liquid interface. The atomized droplets containing Fe^{2+} supplied a medium for the electro-Fenton reaction (*16, 17*), as shown in Eqs (12) and (13). Strong oxidative radicals •OH and $HO_2\bullet$ were produced.

$$Fe^{2+} + H_2O_2 \rightarrow Fe^{3+} + \bullet OH + OH^- \qquad (12)$$

$$Fe^{3+} + H_2O_2 \rightarrow Fe^{2+} + HO_2\bullet + H^+ \qquad (13)$$

In order to determine whether the H_2O_2 passed into the liquid phase, its presence there was monitored in the absence of Fe^{2+} catalyst. The positive result shows that the Fenton reaction was taking place in the atomized droplets.

As previously mentioned, the soluble intermediate products (e.g. HORH•) and H_2O_2 freely dissolved into the Fe^{2+} containing atomized droplets. In the droplets, a further generation of •OH took place in Eq.(12) and the recovery of Fe^{2+} followed Eq.(13). In the presence of •OH, the dissolved products were easily oxidized in reactions Eqs. (7)-(9) and finally into carbon dioxide.

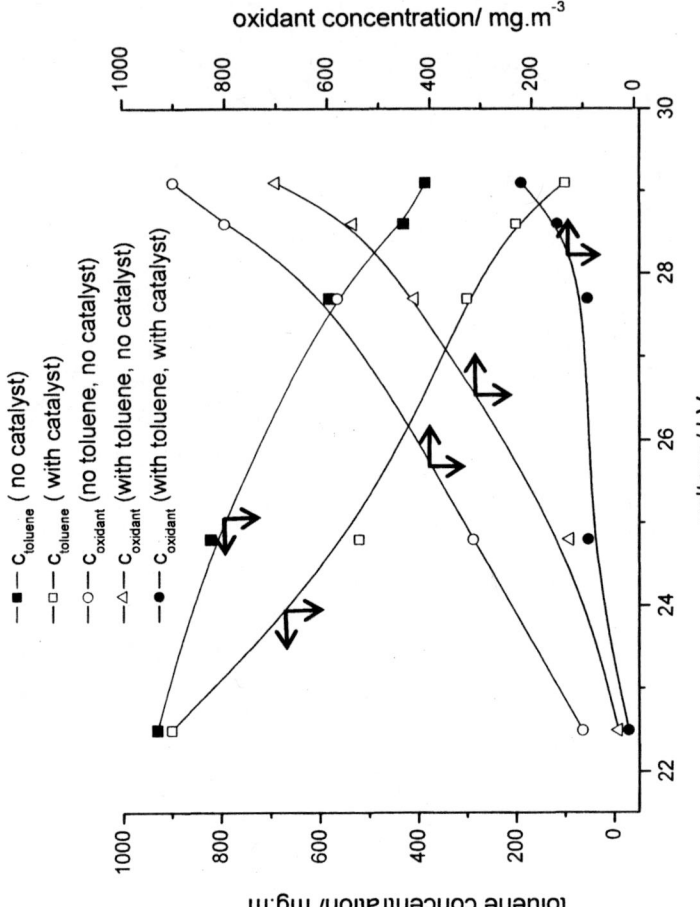

Figure 5. Total oxidant's detection and toluene's decomposition under different applied voltages. Gas flow rate: $0.6\ m^3\ h^{-1}$, toluene initial concentration: $1000\ mg\ m^{-3}$, reaction time: 30 s, electrode gas (O_2): $0.06\ m^3\ h^{-1}$, catalyst: Fe^{2+} 0.05 mM

In the bulk of the solution, there were also many side reactions (18):

$$HO_2\bullet + Fe^{3+} \rightarrow Fe^{2+} + O_2 + H^+ \qquad (14)$$

$$HO_2\bullet + Fe^{2+} \rightarrow Fe^{3+} + HO_2^- \qquad (15)$$

$$HO_2^- + H^+ \rightarrow H_2O_2 \qquad (16)$$

$$HO\bullet + H_2O_2 \rightarrow H_2O + HO_2\bullet \qquad (17)$$

Obviously, in the homogeneous corona radical shower reaction system, the side reactions would compete with the main reaction or consume the oxidants generated. Therefore it is very important to choose suitable concentrations of Fe^{2+} in the catalyst solution and appropriate applied voltages to ensure the main reactions both in the gas phase and in the liquid phase will be given priority.

Effect of catalyst concentration on the decomposition of toluene

Interestingly, different concentrations of Fe^{2+} solutions showed different catalysis ability. A range of concentrations from 0.01-0.2 mmol L^{-1} were added to the catalyst nebulization chamber with the initial concentration of toluene set at c_0 (900 mg m^{-3}) and the reaction time t, was set at 30 s. The higher the applied voltage, the higher the degradation rate ($\overline{v}$) achieved. However, the average degradation rate ($\overline{v}$) and enhancement factor for toluene degradation rate, β (see below), as a function of Fe^{2+} concentration displayed an optimal value between 0.05 mmol L^{-1} and 0.06 mmol L^{-1} (Figure 6). Toluene degradation rate increased from 9.3 mg m^{-3} s^{-1} to 21.8 mg m^{-3} s^{-1} and the enhance factor β increased to 1.34 with the introduction of the 0.05 mmol L^{-1} Fe^{2+} solution. Furthermore, the optimal applied voltage was at 23 kV with minimal value at 29 kV. This may be the reason that when the applied voltage was set above 29 kV, the most intense corona discharge was achieved and the generation rate of active radicals in the gas phase was the highest. It should be noticed that when the Fe^{2+} concentrations were higher than 0.15 mmol L^{-1}, the enhancement was not so obvious, which indicates at that higher concentrations of metallic ions. Other interfering effects may give a reduced reaction rate.

On the other hand, the activation energy was apparently reduced in the presence of the catalyst and the applied voltage needed for the same pollutant degradation was correspondingly reduced. For instance, the maximal degradation rate was 16 mg m^{-3} s^{-1} under an applied voltage of 29 kV without added catalyst. Above the value which the degradation rate could not be increased by raising the applied voltage because spark discharge occurred with

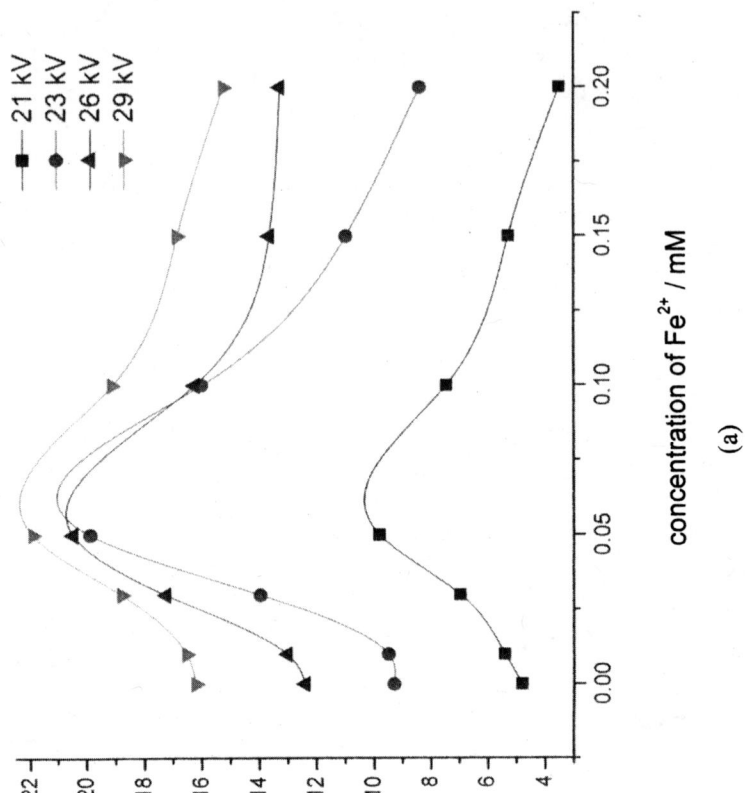

(a)

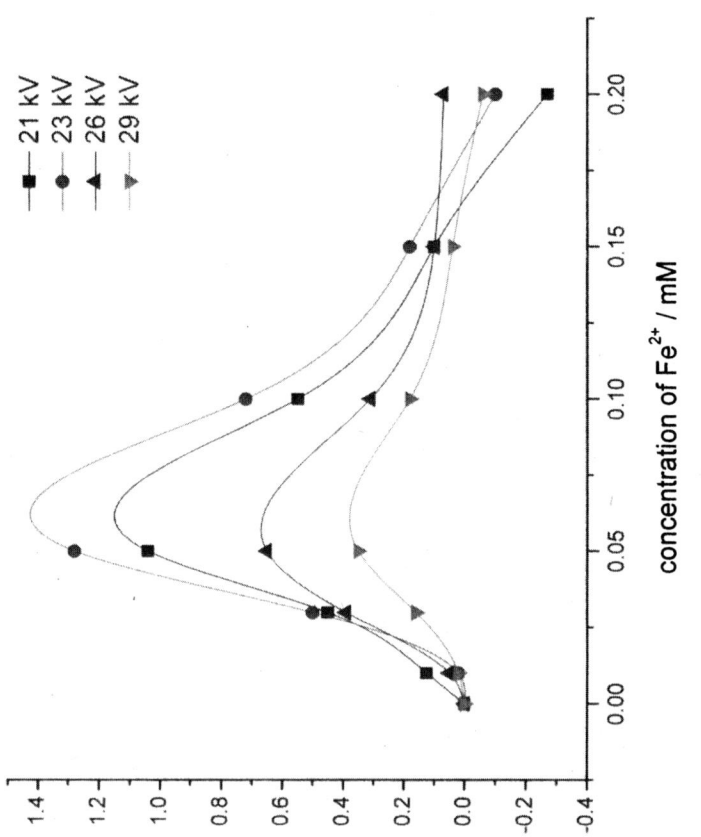

Figure 6. (a) Toluene average degradation rate ν as a function of Fe^{2+} concentration (b) Enhance factor of toluene degradation rate β as a function of Fe^{2+} concentration. Flow rate (mixture of air and toluene): 0.6 $m^3 h^{-1}$, initial concentration: 900 mg m^{-3}, reaction time: 30 s, electrode gas (O_2+H_2O): 0.06 $m^3 h^{-1}$

the electrode design in this experiment. With the addition of 0.05 mmol L^{-1} Fe^{2+} solution into the corona zone during discharge, an applied voltage as low as 22.8 kV was satisfactory to reach the same degradation rate.

Homogeneous catalytic reaction induced by other transition metal ions

Many papers have reported that other low valent transition metal ions, such as Co^{2+} and Mn^{2+}, produced results similar to Fe^{2+} by reacting with H$_2$O$_2$ and produce •OH, (a Fenton-like reaction) (19). Theoretically, these reaction equations occur:

$$Mn^{2+} + 2H_2O_2 \rightarrow Mn^{4+} + 2\bullet OH + 2OH^- \quad (18)$$

$$Mn^{4+} + 2H_2O_2 \rightarrow Mn^{2+} + 2HO_2\bullet + 2H^+ \quad (19)$$

$$Co^{2+} + H_2O_2 \rightarrow Co^{3+} + \bullet OH + OH^- \quad (20)$$

$$Co^{3+} + H_2O_2 \rightarrow Co^{2+} + HO_2\bullet + H^+ \quad (21)$$

A comparison of 0.05 mmol L^{-1} Fe^{2+}, Co^{2+} and Mn^{2+} enhancement of the toluene degradation rate by different applied voltage is displayed in Figure 7. It can be seen that the sequence of the catalytic activity of three catalyst was Mn^{2+} > Co^{2+} > Fe^{2+}. The graph seems to indicate that it will be more promising to focus on Mn^{2+} in the future.

Laboratory experimental studies and approaches

Apparatus and analysis procedures

The corona radical shower system used in the present investigation (Figure 8), includes four parts: corona radical shower reactor, liquid phase Fenton-type catalyst homogenization system, catalyst retrieval and high voltage power supply.

The corona radical shower is induced using a commercially available high voltage power supply variable from 0 to 50 kV. The reactor has a coaxial configuration. A copper tube has nozzles spirally arranged uniformly on the surface, and serves as the discharge electrode, connected to the positive terminal of the power supply. A sheet of stainless steel net covers the inner surface of a PVC cylinder, and forms the grounded electrode, which connects to the negative terminal of the power supply. Positive corona discharge was achieved in this configuration.

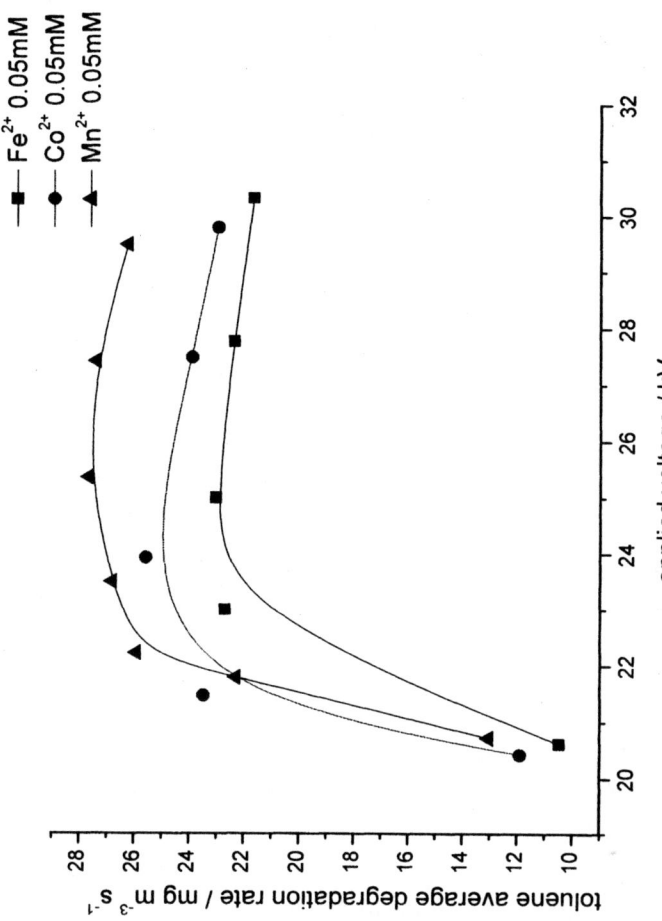

Figure 7. Toluene average degradation rate $\bar{v}$ versus applied voltage by homogenization of heterogeneous catalysis. Flow rate (mixture of air and toluene): $0.6\ m^3\ h^{-1}$, initial concentration: $900\ mg\ m^{-3}$, reaction time: $30\ s$, electrode gas (O_2+H_2O): $0.06\ m^3\ h^{-1}$

Toluene was fed into the reactor from the gas inlet, where it underwent degradation in the corona zone, passing through a catalyst retrieval chamber for deentrainment and finally flowing out through from the gas outlet. O_2, serving as the electrode gas, cames out into the corona zone from the nozzles of the hollow discharge electrode. The experiment was carried out at ambient temperature and atmospheric pressure.

Gas samples were collected at the gas inlet and outlet and liquid samples were collected from the catalyst retrieval chamber. The analyses of toluene in the gas sample were carried out using a gas chromatograph (GC). Indigo disulphonate spectrophotometry was used for detection of O_3 and other oxidants in the gas phase. For the liquid sample, the methods of high performance liquid chromatography (HPLC) and Potassium Titanium(IV) Oxalate spectrophotometry were used to detect the soluble intermediate products and H_2O_2.

The instruments and analytical conditions were: HP 6890 GC (HP, USA) or GC-9790, (Wenling China) with a capillary column (DB-624, 30 m×0.54 mm id, J&W Scientific, USA) equipped with a FID detector; HPLC (Gilson, France), ODS-18 reversed phase column (Alltech, USA), UV detector; 722S spectrophotometer (General Analytical Apparatus Shanghai).

Preparation of the metal ions solution

The desired concentrations of $FeSO_4$ solution with pH 2~3 were prepared and filled into the catalyst nebulizer. Oxygen, used as the electrode gas, went through the catalyst nebulization chamber (shown within the broken line in Figure 1). After nebulization, a mixture of O_2 and innumerable atomized droplets (diameters of 10^{-3} cm to 10^{-7} cm) went into the discharge reactor through the nozzles and formed a homogeneous aerosol.

Degradation rate and enhancement factor

The degradation rate of toluene is defined as:

$$v = -\frac{dc_A}{dt} \tag{22}$$

where, v is the degradation rate of toluene, representing the variation of concentration per unit time under a certain applied voltage with units of $mg\ m^{-3} \cdot s^{-1}$; c_A and c_0 are the final and initial concentrations of toluene in $mg\ m^{-3}$

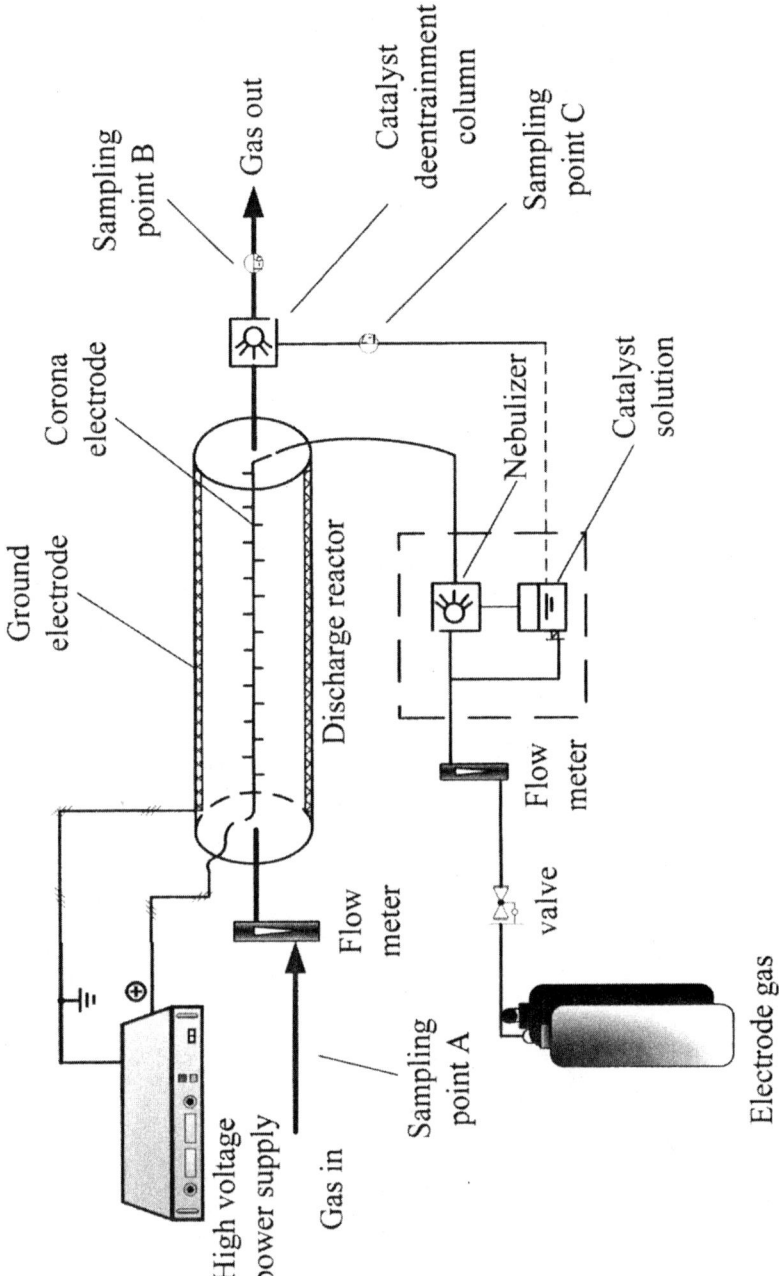

Figure 8. Schematic diagram of experimental system.

respectively and t is the reaction time. When c_0 and t are kept constant, the degradation rate of toluene can be expressed by its average reaction rate $\bar{v}$:

$$\bar{v} = -\frac{c_A - c_0}{t} \tag{23}$$

We define the enhancement factor, β, of toluene degradation in the catalytic reaction as:

$$\beta = \frac{\bar{v}_A - \bar{v}_0}{\bar{v}_0} \tag{24}$$

where, $\bar{v}_A$ is the average degradation rate of VOCs after the addition of the catalyst; $\bar{v}_0$ is the average degradation rate of VOCs in the absence of catalyst.

Conclusions

A mist containing transition metal ions was effective in catalyzing oxidation of toluene by active radicals. The multiphase reactions showed that an insoluble organic substance (tolune) was mainly oxidized in the gas phase and adducts of those soluble organic intermediate products dissolved into the atomized droplets dispersed in the corona radical shower reaction zone. This method involved homogenization of heterogeneous catalysis by dispersing atomized droplets into the gas phase and may have application in air pollution control and in industrial manufacturing processes. Besides Fe^{2+}, other transition metal ions like Mn^{2+} and Co^{2+} also displayed catalytic ability. Further study of the technique will facilitate the development of new technologies to transfom gas-only reactions into gas-mist reactions, thus accelerating their reaction rates.

Acknowledgements

The financial support of the National High-Tech Research and Development Program of China (863 Program) No.2002AA529182 is acknowledged. The authors also thank Richard Dudley for his valuable comments on the manuscript.

Reference

1. FinlaysonPitts, B. J.; Pitts, J. N. *Science.* **1997**, *276*, 1045-1052
2. Roland, U.; Holzer, F.; Kopinke, F. D. *Catal. Today.* **2002**, *73*, 315-323
3. Lee,Y. H.; Jung W. S.; Choi Y. R.; Oh J. S.; Jang, S. D., Son,Y. G.; Cho, M. H.; Namkung, W.; Koh,D. J.; Mok,Y. S.; Chung, J. W. *Environ. Sci. Technol.* **2003**, *37*, 2563-2567
4. Kim,H. H. *Plasma Process. Polym.* **2004**, *1*, 91-110
5. Di, C. P.; Brune, W. H.; Martinez, M.; Harder, H.; Lesher, R.; Ren, X. R.; Thornberry, T.; Carroll, M. A.; Young, V.; Shepson, P. B., Riemer, D.; Apel, E.; Campbell, C. *Science.* **2004**, *304*, 722-725
6. Roland, U.; Holzer, F.; Kopinke, F. D. *Catal. Today.* **2002**, *73*, 315-323
7. Huang, L.; Nakajo, K.; Ozawa, S.; Matsuda, H. *Environ. Sci. Technol.* **2001**, *35*, 1276-1281
8. Nam, T. S. P.; Matthew, V. D. S.; Christopher, W. J.; *Adv. Synth. Catal.* **2006**, *348*, 609-679
9. Torres, R. A.; Petrier, C.; Combet, E.; Moulet, F.; Pulgarin, C. *Environ. Sci. Technol.* **2007**, *41*, 297-302
10. Gromboni, C. F.; Kamogawa, M. Y.; Ferreira, A.. G.; Nobrega, J. A.; Nogueira, A. R. A. *J. Photochem. Photobiol. A-Chem.* **2007**, *185*, 32-37
11. Zhang, L.; Sawell, S.; Moralejo, C.; Anderson, W. A. *Appl. Catal. B-Environ.* **2007**, *71*, 135-142
12. Lee, B. Y.; Park, S. H.; Lee, S. C.; Kang, M.; Choung, S. J. *Catal. Today.* **2004**, *93-95*, 769-776
13. Kang,Y.; Wu, Z. C. *Chin. Sci. Bull.* **2008**, *53*, 1475-1478.
14. Lu, X.; Laroussi, M. *J. Appl. Phys.* **2005**, *98*, 023301
15. Wennberg, P. O.; Hanisco, F.; Jaegle, L.; Jacob, D. J.; Hintsa, E. J.; Lanzendorf, E. J.; Anderson, J. G.; Gao, R. S.; Keim, E. R.; Donnelly, S. G.; DelNegro, L. A.; Fahey, D. W.; McKeen, S. A.; Salawitch, R. J.; Webster, C. R.; May, R. D.; Herman, R. L.; Proffitt, M. H.; Margitan, J. J.; Atlas, E. L.; Schauffler. S. M.; Flocke, F.; McElroy, C. T.; Bui T. P. *Science*, **1998**, *279*, 49-53
16. Faouzi, M.; Canizares, P.; Gadri, A.; Lobato, J.; Nasr, B.; Paz, R.; Rodrigo, M. A.; Saez, C. *Electrochim. Acta.* **2006**, *52*, 325-331
17. Flox, C., Ammar, S.; Arias, C.; Brillas, E.; Vargas-Zavala, A. V.; Abdelhedi, R. *Appl. Catal. B-Environ.* **2006**, *67*, 93-104
18. Tzedakis, T.; Savall,A.; Clifton, M. J. *J. Appl. Electrochem.* **1989**, 19, 911-921
19. Strlic, M., Kolar, J.; Selih, V. S.; Kocar, D.; Pihlar, B. *Acta Chim. Slov.* **2003**, *50*, 619-632

Aerosol Chemistry

Chapter 5

Surface Activity of Perfluorinated Compounds at the Air–Water Interface

Nabilah Rontu and Veronica Vaida

Department of Chemistry and Biochemistry and CIRES, Campus Box 215, University of Colorado, Boulder, CO 80309

The surface activity of two perfluorinated compounds, perfluorododecanoic acid and 8-2 fluorotelomer alchol, was investigated using a Langmuir trough. Surface pressure-area isotherms of the perfluorinated acid and alcohol in single and multi component films containing hydrocarbon organic acids were collected and characterized. Evaporation from the subphase and film stability and miscibility were determined. These results were intended to serve as a proxy for aqueous atmospheric aerosol particles. Our results indicated that both perfluorinated compounds are indeed surface active molecules. In mixed films, perfluorododecanoic acid had the ability to act as both a co-surfactant and had a stabilizing effect. While the 8-2 fluorotelomer alcohol evaoprated from the subphase in pure films, the presence of stearic acid significantly reduced its evaporation and stabilized the film. A consequence of perfluorinated compounds partitioning to the air-water interface is the possibility of their atmospheric transport, distribution, and deposition via aerosol particles.

Introduction

Fully fluorinated organic compounds (perfluorinated compounds) have been used in numerous industrial applications over the past 50 years (*1*). Their unique physical and chemical properties, such as high stability due to their thermal and chemical inertness, low solubilities of long-chain perfluoroalkyl compounds in polar and non-polar solvents, and higher densities and compressibilities as compared to their hydrocarbon analogs, give rise to the various applications of perfluorinated compounds (*2*). The applicability of perfluorinated carboxylic acids, salts, sulfonates, alcohols, and olefins include being used in the synthesis of polymers, surface protectors, and intermediates in products such as surfactants (*3-5*). Some perfluorinated acids have been recently detected in the tissues of animals, in environmental waters, and in precipitation (*6-12*) and other perfluorinated compounds have even been observed in the Arctic food chain (*7, 13*) and in human blood (*14-16*).

Aquatic transport, degradation of volatile precursors, and direct release into the atmosphere are accepted methods of transport for this class of compounds (*4*). In an effort to propose an alternative method of their atmospheric transport, the authors of this work considered the possibility of perfluorinated compounds acting as surfactants on the surfaces of aqueous atmospheric aerosol particles and studied their surface properties (*17, 18*). The results obtained from these studies were intended to serve as a proxy for aqueous atmospheric aerosol particles containing perfluorinated compounds. In the broader context of atmospheric chemistry, the ability of the air-water interface to concentrate and stabilize perfluorinated compounds at the surfaces of aqueous atmospheric aerosol particles has the potential for their long-range transport, distribution, and deposition.

Perfluorinated Compounds as Surfactants

Upon adsorption to substrates, perfluoroalkyl chains repel both water and oil due to their low surface energy, and thus, contribute to perfluorinated compounds being efficient surfactants. Surfactants, by definition, lower the surface tension of a medium by selectively adsorbing on the air-water interface (*19*). Molecules which contain both a hydrophobic tail group and a hydrophilic head group are known as amphiphiles. The antipathy of the hydrophobic tail group for water in combination with the affinity of the hydrophilic head group for water gives rise to surfactant properties in aqueous systems (*20-22*). Amphiphiles have an innate ability to partition to the air-water interface and form self assembled films. In comparison to their hydrocarbon analogs, perfluorinated compounds are much better surfactants because they are able to lower the surface tension at very low concentrations (*4*) and exhibit unique adsorption and partition behavior (*23*).

Atmospheric Implications

In general, perfluorinated compounds have higher vapor pressures than their hydrocarbon analogs and their Henry's law constants and solubilities suggests that these compounds can partition into the air or adsorb onto particles (*4, 24-26*). In the atmosphere, long-range transport via air can occur when substances are in the gas phase or sorbed to a particle or water. Since perfluorinated compounds are well-mixed and long-lived in the troposphere, water solubility, adsorption, and vapor pressure all become important parameters when considering the potential for their long-range transport.

The air-water interface of organic aerosols has the ability to concentrate organics at the surface (*20, 21*). Thus, organics constitute a significant fraction of atmospheric aerosol particle composition (*27-36*), with amphiphiles comprising a large portion of the organic components (*37-39*). An "inverted micelle" structure has been proposed to represent aqueous atmospheric particles (*20, 21*). In this model, an organic film of hydrocarbons, held to the surface by interactions between the polar head groups and the aqueous ionic brine core, coats the aerosol particle, while the hydrophobic tail portions of the molecules are exposed to the highly oxidizing atmosphere where processing can take place. Although perfluorinated acids and alcohols are expected to occur in low concentrations on the surfaces of organic rich atmospheric aerosol particles, a consequence of their partitioning to the air-water interface is the possibility of their transport and widespread distribution and deposition using atmospheric aerosols.

Results and Discussions

The fluorinated compounds used for the study were perfluorododecanoic acid ($CF_3(CF_2)_{10}COOH$, denoted PFDDA), which is the perfluorinated analog of dodecanoic acid, and a partially fluorinated telomer alcohol containing 8 carbons that are fully fluorinated and 2 carbons that are hydrogenated with the formula $CF_3(CF_2)_7CH_2CH_2OH$ (denoted 8-2 FTOH). These compounds were chosen because they are used industrially and also, because they have been observed in the atmosphere.

Experiments were conducted using a Langmuir trough with two mechanical barriers that compress the film. The results of these experiments is a known as a surface pressure-area isotherm which is a plot of surface pressure as a function of the surface area of the surfactants. The shape of the isotherm corresponds to the molecular orientation of the surfactant as a function of barrier position. In the disordered state, the barriers just begin to compress the film and very weak interactions exist between the hydrophobic tails and the subphase surface. As the film is continued to be compressed, the molecules experience more intermolecular attractions with one another and are in closer contact with one another; this is known as the liquid-expanded state. In the liquid-condensed state, the molecules are in rigid contact with one another. Compression beyond this state leads to the phenomenon of collapse where disordered multilayers form. Linear regression of the steep portion of the isotherm (i.e. the liquid-condensed state) to zero surface pressure gives the molecular footprint of the surfactant which is the surface area a single molecule occupies.

Surface pressure-area isotherms of varying concentrations of the compounds in pure and binary mixed films with octadecanoic acid (denoted stearic acid) and dodecanoic acid (denoted lauric acid) were collected. The film properties investigated were stability, miscibility, and evaporation from the subphase. Further details of the experimental procedures and conditions can be found in references 17 and 18.

Perfluorododecanoic Acid

Single Component Systems

Pure systems of three carboxylic acids were studied first to better understand and characterize their individual properties. Figure 1 shows the surface pressure-area isotherms of pure equimolar (10^{-3} M solutions) samples of stearic, lauric, and PFDDA collected at room temperature. The three isotherms have different shapes, collapse pressures, and molecular footprints. As illustrated, the perfluorocarboxylic acid has the largest molecular footprint and collapse pressure.

Stearic acid, with the longest carbon chain length studied here, has a textbook example of an isotherm, clearly showing the disordered (0-0.1 mN/m), liquid-expanded (0.1-27 mN/m), and liquid-condensed phases (27-63 mN/m) (*18*). Linear regression of the steep, liquid-condensed part of its isotherm yields a molecular footprint of 22 $Å^2$/molecule (*18*) which is in excellent agreement with the literature value (*40*) as is the collapse pressure of the film at approximately 60 mN/m (*40*). PFDDA, with six fewer carbons in its backbone than stearic acid, has a molecular footprint of 31.6 $Å^2$/molecule and collapse pressure of approximately 65 mN/m (*18*). The hydrocarbon analog of PFDDA, lauric acid, has a much smaller molecular footprint (16.6 $Å^2$/molecule) and

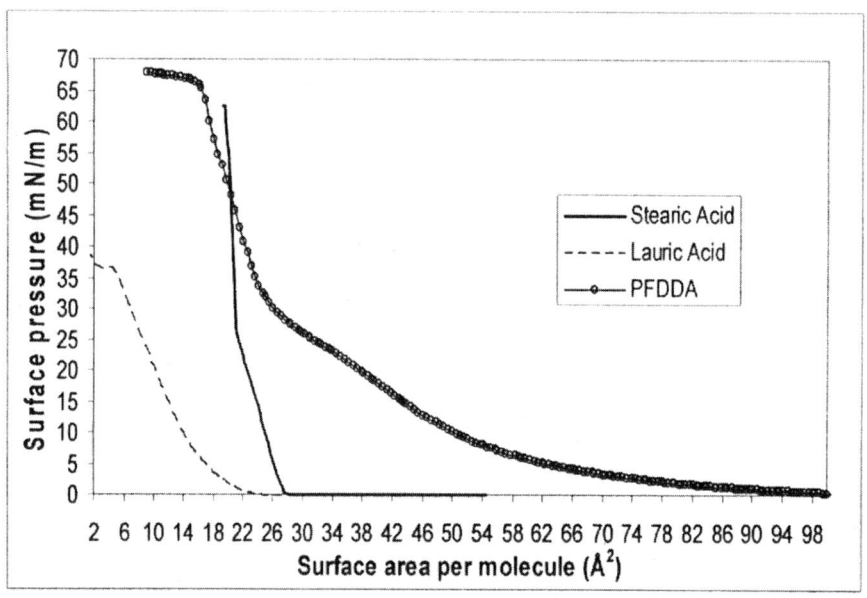

Figure 1. Isotherms of eqimolar samples of stearic acid, lauric acid, and PFDDA. (Reproduced from reference 18. Copyright 2007 American Chemical Society.)

collapse pressure (approximately 37 mN/m), and also, its isotherm is less structured (*18*) than both stearic acid and PFDDA. The larger molecular footprint of PFDDA occurs because fluorocarbons are known to form more expanded monolayers (*41-43*). Since perfluorocarbon chains are highly rigid, apolar and known to interact only by dispersive forces, they experience much smaller van der Waals interaction energy per molecular contact area.

Generally, the hydrophobic tail group determines the efficiency of a surfactant to lower the surface tension. For the PFDDA film, the higher collapse pressure results from the strong electronegative nature of the fluorine atoms (*2, 44*). Also, the low polarizability of perfluorocarbon chains causes weaker interactions with water as compared to that of hydrocarbon chains. As a result, the low cohesive energy per volume of fluorocarbons yields surface tension values which are approximately an order of magnitude smaller than their hydrocarbon analogs (*45*); consequentially, higher surface pressure values are reflected for perfluorocarbons. Thus, in comparison to lauric acid-the hydrocarbon analog of PFDDA, the perfluorocarboxylic acid has a much larger collapse pressure and acts a better surfactant as expected.

Multi Component Systems

The impetus of this work was to understand the behavior of PFDDA in mixtures with other organics since realistically, the surfaces of atmospheric aerosol particles are coated with a mixture of many amphiphiles. Miscibility of PFDDA with stearic and lauric acid was studied by varying its concentration in three combinations in the binary mixtures: 25%, 50%, and 75% PFDDA.

Three different factors were used to determine film miscibility: collapse pressure, the breakpoint K_1 which indicates the phase transition from the liquid expanded to the liquid condensed states, and the additivity rule. First, the collapse pressure is a useful guide to determine film miscibility since a truly miscible film has a single collapse pressure. Film components which are immiscible have well defined and different collapse pressures; isotherms of such immiscible films would show collapse at the lower pressure first, followed by collapse at the higher value. Next, if K_1 is a function of concentration, it suggests complete miscibility between film components at equilibrium (46). Lastly, the additivity rule states that the average area per molecule of any mixture is the sum of the areas occupied by each species at the surface in an ideal behavior. It is given by $A_{avg} = N_1A_1 + N_2A_2$, where A_{avg} is the calculated average area occupied per molecule of the mixed monolayer, N_1 and N_2 represent the corresponding mole fractions of the single components in the mixture, and A_1 and A_2 are the molecular footprints of the pure components. An immiscible film follows the additivity rule, whereas a miscible film deviates from it, indicating that molecular interactions exist between the film components.

Isotherms of the two component mixture of stearic acid and PFDDA were collected which revealed that the film was miscible. Varying the concentration of PFDDA in the mixture did not affect the shape, collapse pressure, or molecular footprint of the isotherms. The general trend showed that the average molecular footprint of the mixture increased as a function of the percent of stearic acid in the mixture. At 25%, 50%, and 75% stearic acid, the average molecular footprints of the mixtures were determined to be 30.6, 31.7, and 33.8 $Å^2$/molecule, respectively (18). Similarly, PFDDA in mixtures with lauric acid produced isotherms which also showed film miscibility. In this case, however, increasing the mole fraction of PFDDA in the mixture generated more structured isotherms with larger molecular footprints. The average molecular footprints of the mixture at 25%, 50%, and 75% lauric acid were shown to be 29.5, 21.1, and 13.1 $Å^2$/molecule, respectively (18). Table 1 shows the experimental and calculated average molecular footprints of the stearic acid/PFDDA and lauric acid/PFDDA mixtures. The last column, A^{Ex}, shows the excess area of mixing, which measures nonideality and is given by $A^{Ex} = A_{act} - A_{avg}$, where A_{act} represents the actual area per molecule of the binary system.

As shown in the table, the mixture of stearic acid with PFDDA shows a positive deviation from ideality upon increasing the percentage of stearic acid in

Table 1. Comparison of Experimental and Calculated Average Molecular Footprints of Mixtures of Stearic Acid and Lauric Acid with PFDDA

	Experimental Footprint ($Å^2$/molecule)	Calculated Footprint ($Å^2$/molecule)	A^{Ex}
Stearic Acid Mixture			
25% PFDDA	33.8	24.4	9.4
50% PFDDA	31.7	26.8	4.9
75% PFDDA	30.6	29.3	1.3
Lauric Acid Mixture			
25% PFDDA	13.1	20.4	-7.3
50% PFDDA	21.1	24.2	-3.1
75% PFDDA	29.5	27.9	1.6

the mixture. This is a result of repulsive interactions between the two components. For the PFDDA/lauric acid mixture, there is a general negative deviation from ideality at larger concentrations of lauric acid in the mixture, which signifies not only an attractive interaction but also miscibility in both the liquid-expanded and liquid-condensed states.

In the mixture with stearic acid, the presence of PFDDA did not greatly affect the structure of the isotherm, the collapse pressure, or the molecular footprint; therefore, it suggests that the PFDDA acts as a co-surfactant in the mixture. However, since increasing the concentration of PFDDA in the mixture with lauric acid gives the isotherms better structure and larger molecular footprints, it implies that the film is stabilized with the presence of PFDDA in the mixture.

8-2 FTOH

Single Component Systems

To be relevant on an atmospheric time scale, surface pressure-area isotherms of 8-2 FTOH were collected at surface layer ages of 5, 10, 30, and 60 minutes to allow for partitioning. Unlike typical fatty acids and alcohols, the isotherms of the fluorotelomer alcohol studied do not show the three distinct phases as can be seen in Figure 2. However, two very important points can be drawn from the figure. First, the isotherms indicate that 8-2 FTOH does partition to the air-water interface, suggesting that the compound is indeed surface active. Second, and more importantly, there is a shift to a smaller molecular footprint and decreased surface pressure with each isotherm as a function of surface layer age. The molecular footprint, which is a characteristic property of an amphiphile and determines how much surface area a single molecule occupies in a compact film at zero pressure, should not change when the concentration and volume are held constant assuming all film components remain at the interface. However, this is not the case as shown in Figure 2. This behavior is attributed to the fact that fluorotelomer alcohols are volatile (*47*) and will evaporate from the surface on these timescales. The possibility that the alcohol could have dissolved into the subphase was ruled out due to the fact that they are only sparingly soluble in water (*47*). Due to evaporation, the isotherms actually show a plot of surface pressure vs. molecular density; thus, the x-axis in Figure 2 is labeled as an uncorrected experimental surface area per molecule.

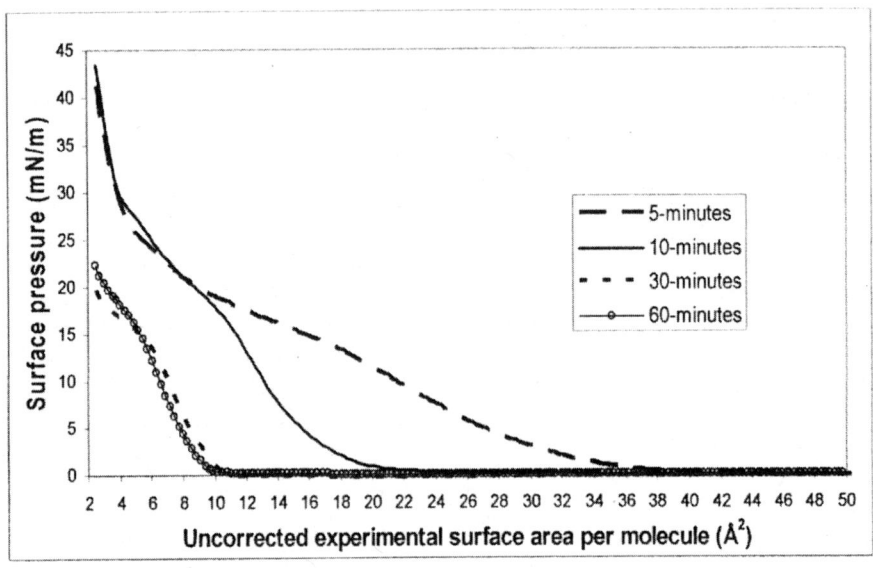

Figure 2. Isotherms of 8-2 FTOH collected at surface layer ages of 5, 10, 30, and 60 minutes. (Reproduced from reference 17. Copyright 2007 American Chemical Society.)

In order to determine the amount of surfactant molecules that remained at the interface as a function of time, it was necessary to establish how much of the fluorotelomer alcohol evaporated. To do this, the molecular footprint of 8-2 FTOH was needed, however, to the best of the authors' knowledge, it has not been previously reported. Due to minimal evaporation, the isotherm collected after 5-minutes was used as a reference and the molecular footprint of 8-2 FTOH was determined to be 34.5 Å^2/molecule which is in agreement with the literature for similar systems (*48*). A correction factor was then applied to the isotherms collected after 10, 30, and 60-minutes to account for evaporation. From this data, the evaporation rate of 8-2 FTOH was established at 0.384% (± 0.002%) per minute (*17*). Further explanations detailing the methods used to verify the evaporation rate can be found in reference 17.

Multi Component Systems

The fluorotelomer alcohol was also studied in a binary one-to-one molar mixture with stearic acid to investigate its behavior in mixed systems. For consistency with the pure 8-2 FTOH data, experiments were carried out in the same manner using identical volumes, times, and concentrations. Stearic acid was chosen because it is stable at the interface over long periods of time. Figure 3 shows the isotherms of the binary mixture at various surface layer ages. There are no shifts towards decreased molecular footprints and surface pressures with time, unlike in the case of the isotherms of the pure fluorotelomer alcohol. Isotherms of the mixed film retain similar shapes, surface pressures, and limiting

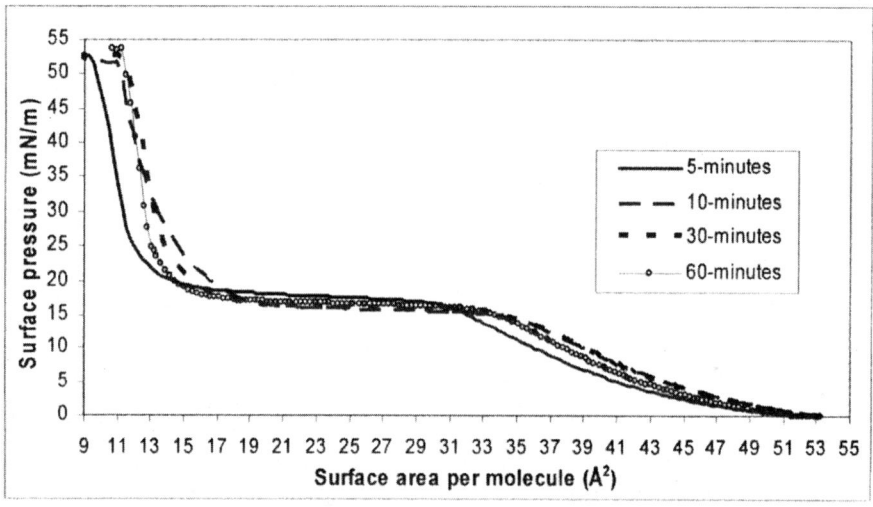

Figure 3. Isotherms of 8-2 FTOH and stearic acid in a binary equimolar mixture at surface layer ages of 5, 10, 30, and 60-minutes. (Reproduced from reference 17. Copyright 2007 American Chemical Society.)

molecular areas regardless of the surface layer age on the subphase at the times investigated here.

Addition of stearic acid to the fluorotelomer alcohol, regardless of time, consistently formed stable monolayers and significantly reduced evaporation as compared to the pure 8-2 FTOH film. Previous studies have shown that stearic acid helps to form highly stable, compressible monolayer when mixed with surfactants incapable of forming stable monolayer at the air-water interface (49, 50). The outcome of film stability has several consequences in the atmosphere. First, if the fluorotelomer alcohol is present on the surfaces of atmospheric aerosol particles with other organics, it has the potential to not only partition to, but also remain at the air-water interface. Its stability in mixed films makes atmospheric aerosol particles plausible candidates for transporting and distributing this class of compounds in the atmosphere. Also, being present on atmospheric aerosol surfaces can lead to further heterogeneous chemistry of fluorotelomer alcohols, which can possibly be very different from reactions expected to occur in the gas phase, resulting in important atmospheric consequences.

The isotherms of the binary mixture also suggest that the two components form a film that is immiscible. Two distinctly different collapse pressures are shown in Figure 3, at 20 and 55 mN/m, which is a textbook example of immiscibility. While 8-2 FTOH did not collapse on its own, the presence of stearic acid helps to stabilize the fluorotelomer alcohol. Differences in the cohesive forces that exist between 8-2 FTOH and stearic acid prevent the components from mixing well and form islands/aggregates when mixed together. A recent study found that the liquid phase of aerosols is highly concentrated with organics and offers a reaction medium very different from aqueous solutions of fog or rain droplets (51). The surfaces of atmospheric aerosol particles are coated with a mixture of many organics, indicating that if the fluorotelomer alcohol is present in the mixture, it has the potential to partition to and remain at the interface.

Conclusion

Recent field measurements have shown that organics comprise a significant fraction of atmospheric aerosol particles. The detection of perfluorinated compounds in environmental waters and the atmosphere was the impetus for this research since not much data are available regarding the atmospheric transport for this class of compounds. Although much work has been done on understanding the gas phase chemistry of perfluorinated compounds, its heterogeneous chemistry has not been explored to the same depth.

In this study, films of perfluorododecanoic acid (PFDDA) and 8-2 FTOH were studied at the air-water interface using a Langmuir trough as a representative model of atmospheric aerosol particles. PFDDA acid was verified to be a better surfactant than its hydrocarbon analog acid. In mixed films, PFDDA has the ability to act as a co-surfactant and have stabilizing effects. The surface activity of 8-2 FTOH was also studied and confirmed. Evaporation of the pure fluorotelomer alcohol was significantly reduced in the presence of stearic acid. Mixed films are better representations of atmospheric aerosol particles, and if perfluorinated compounds partition to and remain at the air-water interface like the data suggests, further chemistry is possible. Results of this work suggests that aerosol particles are a legitimate candidate for the transport, distribution, and deposition of perfluorinated compounds.

References

1. Banks, R. E.; Tatlow, J. C. Organofluorine Chemistry: Nomenclature and Historical Landmarks. *In Organofluorine Chemistry: Principles and Commercial Applications*; Banks, R. E., Smart, B. E., Tatlow, J. C., Eds; Plenum Press: New York, 1994; pp 1-21.
2. Lehmler, H. J.; Oyewumi, M. O.; Jay, M.; Bummer, P. M. *J. Fluorine Chem.* **2001**, *107*, 141146.
3. Hori, H.; Takano, Y.; Koike, K.; Kutsuna, S.; Einaga, H.; Ibusuki, T. *Appl. Catal. B: Environ.* **2003**, *46*, 333.
4. Prevedouros, K.; Cousins, I. T.; Buck, R. C.; Korzeniowski, S. H. *Environ. Sci. Technol.* **2006**, *40*, 32-44.
5. Kissa, E. *Fluorinated Surfactants and Repellents, 2nd ed.; Surfactant Science Series*; Marcel Dekker: New York, 2001; Vol. 97.
6. Shine, K. P.; Gohar, L. K.; Hurley, M. D.; Marston, G.; Martin, D.; Simmonds, P. G.; Wallington, T. J.; Watkins, M. *Atmos. Environ.* **2005**, *39*, 1759-1763.
7. Kannan, K.; Yun, S. H.; Evans, T. J. *Environ. Sci. Technol.* **2005**, *39*, 9057-9063.
8. Hori, H.; Hayakawa, E.; Einaga, H.; Kutsuna, S.; Koke, K.; Ibusuki, T.; Kiatagawa, H.; Arakawa, R. *Environ. Sci. Technol.* **2004**, *38*, 6118-6124.
9. Schultz, M. M.; Barofsky, D. F.; Field, J. A. *Environ. Sci. Technol.* **2006**, *40*, 289-295.
10. Scott, B. F.; Moody, C. A.; Spencer, C.; Small, J. M.; Muir, D. C. G.; Mabury, S. *Environ. Sci. Technol.* **2006**, *40*, 6405-6410.
11. Giesy, J. P.; Kannan, K. *Environ. Sci. Technol.* **2002**, *36*, 146A.
12. Martin, J. W.; Smithwick, M. M.; Braune, B. M.; Hoekstra, P. F.; Muir, D. C. G.; Mabury, S. A. *Environ. Sci. Technol.* **2004**, *38*, 373-380.

13. Smithwick, M.; Muir, D. C. G.; Mabury, S. A.; Solomon, K. R.; Martin, J. W.; Sonne, C.; Born, E. W.; Letcher, R. J.; Dietz, R. *Environ. Toxicol. Chem.* **2005**, *24*, 981-986.
14. Hansen, K. J.; Clemen, L. A.; Ellefson, M. E.; Johnson, H. O. *Environ. Sci. Technol.* **2001**, 35, 766-770.
15. Kannan, K.; Corsolini, S.; Falandysz, J.; Fillmann, G.; Kumar, K. S.; Loganthan, B. G.; Mohd, M. A.; Olivero, J.; van Wouwe, N.; Yang, J. H.; Aldous, K. M. *Environ. Sci. Technol.* **2004**, *38*, 4489-4495.
16. Taniyasu, S.; Kannan, K.; Hori, Y.; Yamashita, N. *Environ. Sci. Technol.* **2004**, *37*, 2634-2639.
17. Rontu, N.; Vaida, V. *J. Phys. Chem. C* **2007**, *111*, 11612-11618.
18. Rontu, N.; Vaida, V. *J. Phys. Chem. C* **2007**, *111*, 9975-9980.
19. Donaldson, D.J.; Vaida, V. *Chem. Rev.* **2006**, *106*, 1445-1461.
20. Gill, P. S.; Graedel, T. E.; Weschler, C. *J. Rev. Geophys.* **1983**, *21*, 903-920.
21. Ellison, G. B.; Tuck, A. F.; Vaida, V. *J. Geophys. Res. Atmos.* **1999**, *104* (D9), 11633-11641.
22. Dobson, C. M.; Ellison, G. B.; Tuck, A. F.; Vaida, V. *Proc. Natl. Acad. Sci. U.S.A.* **2000**, *97*, 11864.
23. Goss, K.-U.; Bronner, G. *J. Phys. Chem. A* **2006**, *110*, 9518-9522.
24. Dinglasan, M. J. A.; Ye, Y.; Edwards, E. A.; Mabury, S. A. *Environ. Sci. Technol.* **2004**, *38*, 2857-2864.
25. Goss, K.-U.; Bronner, G.; Harner, T.; Hertel, M.; Schmidt, T. C. *Environ. Sci. Technol.* **2006**, *40*, 3572-3577.
26. Krusic, P. J.; Marchione, A. A.; Davidson, F.; Kaiser, M. A.; Kao, C. P. C.; Richardson, R. E.; Botelho, M.; Waterland, R. L.; Buck, R. C. *J. Phys. Chem. A* **2005**, *109*, 6232-6241.
27. Mayol-Bracero, O. L.; Gabriel, R.; Andreae, M. O.; Kirchstetter, T. W.; Novakov, T.; Ogren, J.; Sheridan, P.; Streets, D. G. *J. Geophys. Res. Atmos.* **2002**, 107 (D19), INX229/1-INX229/21.
28. Allan, J. D.; Alfarra, M. R.; Bower, K. N.; Williams, P. I; Gallagher, M. W.; Jimenez, J. L.; McDonald, A. G.; Nemitz, E.; Canagaranta, M. R.; Jayne, J. T.; Coe, H.; Worsnop, D. R.; *J. Geophys. Res. Atmos.* **2003**, 108 (D3), AAC2/1-AAC2/15.
29. Middlebrook, A. M.; Murphy, D. M.; Thomson, D. S. *J. Geophys. Res. Atmos.* **1998**, 103 (D13), 16475-16492.
30. Murphy, D. M.; Thomson, D. S.; Mahoney, T. M. J. *Science* **1998**, 282, 1664-1669.
31. Murphy, D. M.; Thomson, D. S.; Middlebrook, A. M.; Schein, M. E. *J. Geophys. Res. Atmos.* **1998**, 103 (D13), 16485-16492.
32. Novakov, T.; Corrigan, C. E.; Penner, J. E.; Chuang, C. C.; Rosario, O.; Bracero, O. L. *J. Geophys. Res. Atmos.* **1997**, 102 (D17), 21307-21314.

33. Novakov, T.; Penner, J. E. *Nature* **1993**, 365, 823-826.
34. Tervahattu, H.; Hartonen, K.; Kerminen, V. M.; Kupiainen, K.; Aarnio, P.; Koskentalo, T.; Tuck, A. F.; Vaida, V. *J. Geophys. Res. Atmos.* **2002**, 107 (D7), AAC1/1-AAC1/8.
35. Tervahattu, H.; Juhanoja, J.; Kupiainen, K. *J. Geophys. Res. Atmos.* **2002**, 107 (D16), ACH18/1-ACH18/7.
36. Tervahattu, H.; Juhanoja, J.; Vaida, V.; Tuck, A. F.; Niemi, J. V.; Kupiainen, K.; Kulmala, M.; Vehkamaki, H. *J. Geophys. Res. Atmos.* **2005**, 110 (D6), D06207/1-D06207/9).
37. Alves, C.; Carvalho, A.; Pio, C. *J. Geophys. Res. Atmos.* **2002**, 107 (D21), ICC7/1 ICC7/9.
38. Fraser, M. P.; Cass, G. R.; Simoneit, B. R. T. *Environ. Sci. Technol.* **2003**, 37, 446 453.
39. Mochida, M.; Kawamura, K.; Umemoto, N.; Kobayashi, M.; Matsunaga, S.; Lim, H. J.; Turpin, B. J.; Bates, T. S.; Simoneit, B. R. T. *J. Geophys. Res. Atmos.* **2003**, 108 (D23), ACE6/1-ACE6/12.
40. Roberts, G. *Langmuir-Blodgett Films*; Plenum Press: New York,1990.
41. Gaines, G. L. *Langmuir* **1991**, 7, 3054-3056.
42. Shafrin, E. G.; Zisman, W. A. *J. Phys. Chem.* **1957**, 61, 1046-1053.
43. Fox, H. W. *J. Phys. Chem.* **1957**, 61, 1058-1062.
44. Rolland, J. P.; Santaella, C.; Vierling, P. Chem. Phys. Lipids 1996, 79, 71.
45. Fletcher, P. D. I. *Fluorinated and semi-fluorinated surfactants. In Specialist Surfactants*; Robb, I. D., Ed.; Blackie Academic and Professional: New York, 1997; pp 104-143.
46. Dorfler, H. *Adv. Colloid Interface Sci.* **1990**, 31, 1-110.
47. Kaiser, M. A.; Cobranchi, D. P.; Kao, C. P. C.; Krusic, P. J.; Marchione, A. A.; Buck, R. C. *Chem. Eng. Data* **2004**, 49, 912-916.
48. Lehmler, H. J.; Bummer, P. M. *J. Fluorine Chem.* **2002**, 117, 17-22.
49. Gilman, J. B.; Eliason, T. L.; Fast, A.; Vaida, V. *J. Colloid Interface Sci.* **2004**, 280, 234-243.
50. Hussain, S. A.; Deb, S.; Bhattacharjee, D. *J. Lumin.* **2005**, 114, 197-206.
51. Marcolli, C.; Luo, B.P.; Peter, T. *J. Phys. Chem. A* **2004**, 108, 2216-2224.

Chapter 6

Atmospheric Chemistry of Urban Surface Films

D. J. Donaldson[1], T. F. Kahan[1], N. O. A. Kwamena[1,2], S. R. Handley[1], and C. Barbier[1]

[1]Department of Chemistry, University of Toronto, 80 St. George Street, Toronto, Ontario M5S 3HS, Canada
[2]Current address: School of Chemistry, University of Bristol, Bristol BS8 1TS, United Kingdom

> The heterogeneous chemistry taking place in and on the films which coat impervious urban surfaces has received almost no attention to date. The chemical composition of such surface films suggests that heterogeneous oxidation reactions and photochemical transformations could play important roles in the environment, potentially affecting local air quality. In this paper we briefly review our work in this area.

Introduction

Heterogeneous atmospheric chemistry has predominantly been concerned with processes involving airborne aerosol particles. This is mainly for historical reasons: acid rain and ozone hole (and depletion, generally) chemistry has demanded an understanding of chemical processes taking place on and at the surface of such particles. *(1)* More recently, the interest has also been motivated by trying to understand and quantify the cloud-condensing properties of aerosols, and how these properties may change as the particles are aged in the atmosphere. However, it is certainly true that similar chemical interactions may take place among gas-phase species and other heterogeneous media, such as bodies of water, vegetation and soils. In urban environments there also exists the possibility of reactions taking place on the large and stationary surfaces of the films coating buildings, roadways, etc. Organic films (self-assembled monolayers or neat liquids) have been used as laboratory proxies for aerosol surfaces *(2)* because of experimental challenges often associated with exploring reactions using "real" particles. However, reactions on urban films could display rather different features than the corresponding processes in aerosol, gas or aqueous phases, due to the different chemical environment of the exposed urban surfaces. Yet, there are few explorations of the chemistry taking place on such surfaces. The following is a brief review of our recent work in this direction.

The Nature of Urban Surface Films

In the past decade, extensive work by Diamond and co-workers *(3-7)* has shown that exposed outdoor surfaces in urban areas become rapidly coated with a complex mixture of chemical compounds ("urban surface film"), most readily encountered as "window grime." This film grows via accretion from the atmosphere and is removed by rain washoff, or revolatilization processes, yielding an (estimated) steady-state thickness of several tens to hundreds of nanometers. Chemical analysis of these films has been carried out, both in a "broad brush" approach, *(6, 7)* which identified the compound classes present, and more detailed studies, *(3-5)* that determined the specific compounds within these classes.

Interestingly, organics make up only 5-10% (by mass) of the films; most of the identified mass is nitrate (~7%), sulphate (~8%) and various metals (18%). Among the organics, fatty acids, alkanes, carbohydrates and aromatics are all observed, as well as trace contaminants such as PAHs, PCBs and PBDEs. Field measurements have determined that, for the most part, the partitioning of such trace gases between the atmosphere and the film is related to the octanol-air partitioning coefficient, K_{OA}, *(4)* suggesting that it is the organic fraction which controls the deposition of airborne compounds to the film. This property has been exploited by using the urban films as passive samples of ambient air pollutants. *(3, 8, 9)* From a laboratory perspective, it suggests that proxy films composed of octanol, mixed with other components, might mimic the chemical environment of "real" urban films.

Chemical Reactions on Urban Films

Experimental Studies

At this time, there have been no laboratory studies reported investigating reactions on "real" urban films, i.e. films allowed to develop outdoors in an urban setting. Most studies have involved oxidation of single-component (e.g.. self-assembled monolayers, oleic acid or solid PAH – see Rudich *(2)* for a recent review) organic films by ozone, though some very recent work on OH and NO_3 has been reported. *(10-13)* In all cases studied to date, the reaction rate is significantly faster than the corresponding gas-phase value.

We have recently reported on an extensive study of the reaction kinetics of gas-phase ozone with six PAH compounds embedded in organic films composed of octanol or decanol, mixed with vacuum grease and (for some experiments) other organic compounds. *(14)* Octanol was chosen as the primary proxy for the "true" film, because of the strong relationship between the K_{OA} and K_{FA} values for the compounds chosen here, as discussed above. Our studies have used laser-induced fluorescence detection of the PAH compound to track its concentration as a function of time following exposure to ozone. The full results are given in Kahan et al. *(14)* and shown in Tables I and II; some key points are summarized here.

Table I. Heterogeneous reaction rates for PAHs in organic films with gas-phase ozone ([O$_3$] =50 ppb)

PAH	k_{obs} (10^{-7} s^{-1})
Benzo[*a*]pyrene	28.3
Anthracene	16.5
Naphthalene	5.93
Pyrene	4.61
Phenanthrene	2.64
Fluoranthene	≤ 0.92

The first important result is that the reactions appear to take place at the surface of the organic film. This is indicated by the non-linear dependence of the measured pseudo-first order rate coefficients on the gas-phase ozone concentration, which show a Langmuir–Hinshelwood form in all cases. Figures 1 and 2 illustrate the kinetic results for reaction with benzo[*a*]pyrene. The data shown in Figure 2 are well fit to a Langmuir–Hinshelwood model, which implies that the ozone reagent undergoes equilibrium (or, at least steady-state) partitioning between the gas phase and the surface and that the reaction takes place between surface-bound ozone and surface adsorbed PAH. Consistent with this we found that anthracene is surface active in octanol *(14)* (although not nearly so much as in water *(15)*), confirming the availability of the PAHs for reaction at the air-organic film interface.

The finding that the reaction of ozone with PAHs takes place at the organic film surface was unexpected, given the solubilities of the compounds in octanol. We and others have previously reported the Langmuir–Hinshelwood kinetic mechanism for such ozonation reactions on water *(16-19)* and solid surfaces *(20-23)*. This surface reactivity has also very recently been found on liquid

Table II. Effect of composition of organic film on anthracene kinetics ($[O_3]$ = 50 ppb)

Substrate	k_{obs} (10^{-7} s^{-1})*
Quartz Plate	18.0
Silicone Grease	16.3
Vacuum Grease	17.2
Decanol	16.2
Corn Starch	20.5
Stearic Acid	23.5
Oleic Acid (1:1 v/v%)	5.6
Squalene (1:1 v/v%)	1.8

*the error in the k_{obs} is ±25%

aerosol particles composed of organic substrates (unreactive towards ozone), *(21)* and so appears to be a general feature of these reactions. This has important consequences for how such reactions are treated in atmospheric chemistry models – whether it is the exposed surface area or the available volume of substrate which is considered.

The second key result is that the presence of other compounds did not affect the observed reaction kinetics, except when the octanol solvent was mixed with an equal volume of oleic acid. Here, the loss rate of anthracene was diminished by about 70%, perhaps due to scavenging of ozone by the highly reactive fatty acid, or to the formation of a polymer "shell" by radical reactions involving the oleic acid and its reaction products with ozone. However, the independence of the loss rates on the presence of other compounds, representative of the classes identified in the organic portion of urban surface films, suggests that laboratory experiments on simplified film proxies could yield results which have relevance to real situations. Some modeling results which address the importance of such processes are discussed below.

Modelling Studies

We incorporated the ozonation kinetic results outlined above into a modified multimedia partitioning model, in order to estimate the overall importance of surface reactions to the fate of PAHs in an urban setting. The MUM-Fate model, developed by Diamond and co-workers, *(24, 25)* is based on

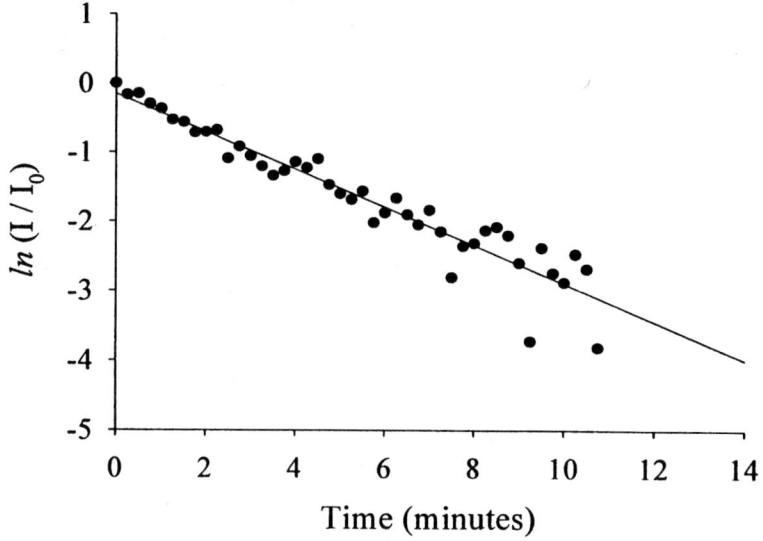

Figure 1. Loss kinetics of benzo[a]pyrene in an organic film due to reaction with gas-phase ozone.

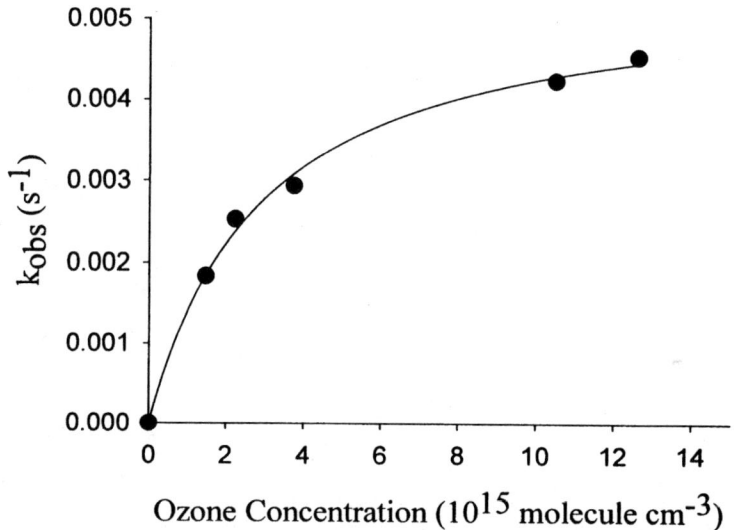

Figure 2. Dependence of the measured first-order rate coefficients for heterogeneous benzo[a]pyrene loss on the gas-phase ozone concentration. The curve shows a fit to a Langmuir–Hinshelwood kinetic model.

the level III steady-state fugacity model of Mackay. *(26)* It consists of seven media compartments: lower air from 0-50 m, which contains aerosol particles; upper air of 50-500 m, which also contains aerosol particles; surface soil; surface water; sediment; vegetation and the film on impervious surfaces. The six PAHs whose heterogeneous ozonation kinetics were studied by us were allowed to enter the model through direct emission into the lower air compartment. Chemical transfer between compartments is characterized by chemical- and media-specific transport parameters. Chemical loss from the model urban system occurs through advective flows from air and water, reactions in all media, burial from sediment, and leaching from soil. Full details are given in Kwamena et al. *(27)*

Several different model environments with differing amounts of impervious surface area (and hence surface film) were studied. The model results suggest that the reactive fate of PAHs in the urban environment is determined by the interplay between their mass distribution, governed by partitioning out of the air compartment, and their lifetimes in the different environmental compartments. Advection was the dominant loss process in all model environments for all PAHs investigated. However, films coating impervious surfaces were found to play an important role for the heavier, lower volatility PAHs, which accumulate more easily in organic condensed phases. For one such compound, benzo[*a*]pyrene, ~75% of the chemical loss in highly urbanized model environments occurred via film reaction; amounting to ~30% of the total loss (i.e. considering both advective and reactive loss process) under stagnant air conditions. The more volatile PAHs remained almost entirely in the air compartment, and so were minimally affected by reactions on urban films; their chemical loss fate was dominated by gas-phase reactions with OH.

These model results strongly suggest that under certain environmental conditions, heterogeneous reactions on urban films may play an important role in determining the fates of low volatility PAHs. Although most of the PAH mass was exported from the model urban environment, the film reactions give rise to products which are also associated with the films. These products are probably more highly oxidized than the parent compounds, and hence likely to remain in the condensed phase.

It should be noted that the only heterogeneous reactions included in the model were those of the PAHs with ozone, for which the kinetic parameters were available. There have been only a handful of studies reported of the reactions of other atmospheric oxidants with organic films. In almost all cases, the films were composed of a single organic compound (either liquid or a self-assembled monolayer). Under these conditions, the reaction rates were found to be quite high, with reaction probabilities towards OH typically > 10% per collision with the organic film surface. *(10)* Reaction efficiencies towards NO_3 radical seem to be lower, *(11, 12)* but its high nighttime abundance suggests that it may be a key oxidant of compounds in such films. As more such reactions are studied experimentally, it will be important to revisit this topic, in order to determine a more quantitative picture of the fates of semivolatile organics in an urban setting.

Photochemistry of Compounds Associated with Urban Organic Films

Given the rich chemical composition of urban surface films, it would be surprising if there were not important photochemical reactions taking place within them. Indeed, the appearance of polymeric compounds within the film is certainly consistent with photochemically-driven reactions, though the origin of the polymers has not yet been established. We have started to study the photochemistry taking place within such films, concentrating on the role(s) which inorganic nitrate anions might play.

In one study *(28)* we deposited nitric acid from the gas phase onto films prepared in a similar manner to those described above. *(14)* Using acridine, a pH-sensitive fluorescent probe, we observed acidification of the film upon exposure to $HNO_3(g)$, indicating that the acid was taken up by the organic film and remained there in (at least partially) dissociated form. Illumination of this acidified film using the output of a Xe lamp, filtered optically to simulate actinic radiation on Earth's surface, caused the pH to increase, eventually returning to its original value (i.e. – that it displayed before acidification). Figure 3 displays these changes in the emission spectrum. At the same time, the concentration of nitrate anion also diminished, as measured by ion chromatography. Given the known photochemistry of nitrate anion in water and ice, *(29-33)* and other arguments presented in Handley et al., *(28)* we proposed that these observations indicate that the nitrate anion in organic films could photochemically generate NO_2 and HONO, which are then released to the gas phase.

This could have important atmospheric consequences. Because the primary pathway for removal of inorganic nitrate (nitric acid or ammonium nitrate) from the atmosphere is by wet (i.e. uptake by water droplets) or dry deposition, followed by rainout / wash off to the ground, this photochemical reduction of NO_3^- provides a mechanism to recycle nitrate back to the gas phase as "active" nitrogen oxides (HONO, NO_2 or NO). A simple calculation of the steady-state lifetime of nitrate (assuming 7% by mass) in the urban films yields ~ 3 days; the average lifetime against rainoff in Toronto is ~ 8 days. *(34)* This suggests other mechanisms for removal of nitrate from the film, consistent with the photochemical mechanism proposed by us. *(28)*

Another way that sunlight can influence the chemistry in the films is via photolysis of the organic compounds present there. We have studied the direct and indirect photochemical loss rates of three PAH compounds in octanol solution, to start to ascertain the importance of this pathway. Solutions of anthracene and pyrene in octanol (~ 10^{-4} mol L^{-1}) were prepared, placed in 1 cm pathlength cuvettes, then illuminated using the optically filtered output of a 100 W Xe arc lamp. The rate of decay of the PAH compounds was determined by measuring resolved fluorescence spectra after fixed irradiation times. As illustrated in Figure 4, in all cases good fits to single exponential decays were obtained; much poorer fits were found assuming second-order kinetics.

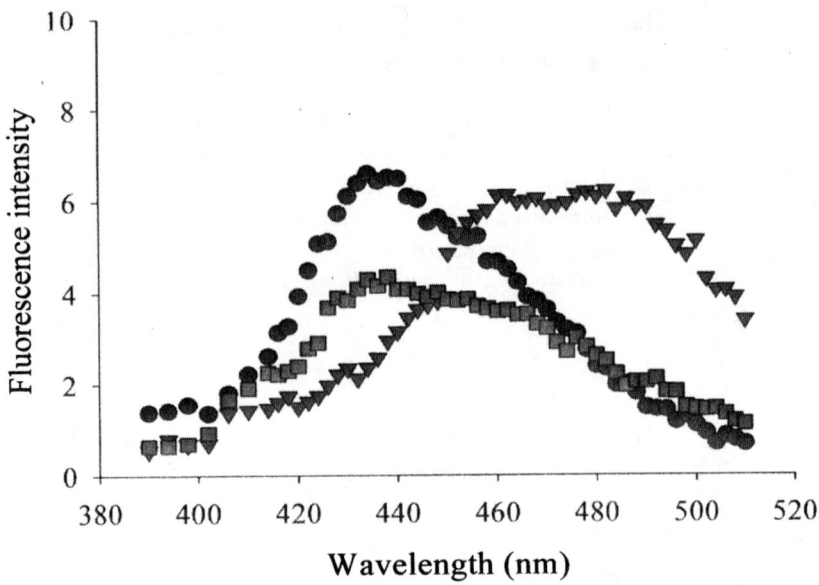

Figure 3. Fluorescence of acridine in organic film before exposure to gas-phase nitric acid (circles); following exposure to gas-phase acid (triangles); and following 90 minutes of irradiation with actinic light (squares).

Actinometry was not performed in these experiments, so the absolute decay rates cannot be related to rates under actinic illumination. However, the experiments were carried out in tandem with identical experiments using water as solvent, so the loss rates relative to that in water could be established. The photolysis rates in octanol using radiation of $\lambda > 305$ nm were 5-10 times lower than the corresponding rates in aqueous solution. For anthracene, the photolysis rate decreased from $(2.5\pm0.3)\times10^{-4}$ s^{-1} in water to $(2.4\pm0.9)\times10^{-5}$ s^{-1} in octanol; that of pyrene decreased from $(1.8\pm0.5)\times10^{-4}$ s^{-1} to $(2.3\pm0.7)\times10^{-5}$ s^{-1}. Photolysis lifetimes of these PAHs in water have been measured using actinic radiation; there is a lot of scatter in the data (see Mackay et al. *(35)* for a listing), but generally they are on the order of several hours, suggesting that their lifetimes in organic films could be on the order of several hours to a day.

An interesting effect was noted when nitrate was also present in solution. In the presence of 10^{-3} mol L^{-1} ammonium nitrate, an enhancement of about 50% in the photolysis rate was observed for each of the PAHs in both water and octanol solution. In aqueous solution, we have measured a linear dependence of the photolysis rate of anthracene on the concentration of nitrate anion; at the

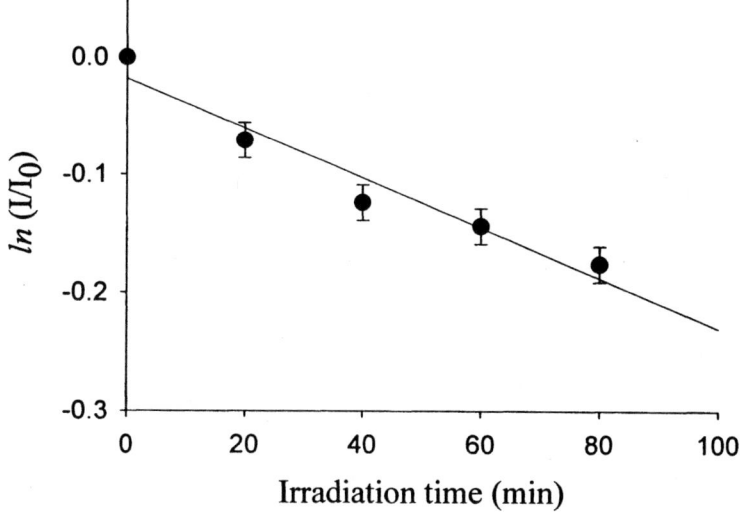

Figure 4. Decay kinetics of anthracene in octanol, under illumination from 100 W Xe lamp with a 300 nm long-pass cutoff optical filter. The good fit to a straight line indicates first-order loss kinetics.

same time, we observe an increase in the initial yield of anthraquinone, the only product detected. Since the enhancement factors of the photolysis are the same in the two solvents, it is likely that their mechanisms are similar as well. In water, nitrate enhances the formation of anthraquinone, presumably via production of OH by nitrate photolysis and that product's reaction with anthracene. If this is true as well in urban films, where there is a high concentration of nitrate, such reactions could be responsible for forming some of the oxygenated aromatics present there.

Conclusions

There have been only a small number of studies to date of the chemistry taking place in or on the "urban grime film," that coats urban impervious surfaces. However, to date indications are that such chemistry could be important in determining local oxidative strength (through photochemical NO_x recycling) and pollutant concentrations. Clearly much work remains to establish the relative importance of different gas phase oxidants, to determine reaction products, and to include such processes in models for urban air quality.

Acknowledgements

The work described herein was funded by NSERC and CFCAS. We have benefited from many years of interesting discussions with Prof. Miriam Diamond concerning urban films.

References

1. Finlayson-Pitts, B.J. and Pitts, J.N., *Chemistry of the Upper and Lower Atmosphere*. 2000, San Diego: Academic Press. 969.
2. Rudich, Y., *Chem. Rev.*, **2003**, *103*, 5097-5124.
3. Butt, C.M., Diamond, M.L., Truong, J., Ikonomou, M.G., Helm, P.A., and Stern, G.A., *Environ. Sci. Technol.*, **2004**, *38*, 3514-3524.
4. Diamond, M.L., Gingrich, S.E., Fertuck, K., McCarry, B.E., Stern, G.A., Billeck, B., Grift, B., Brooker, D., and Yager, T.D., *Environ. Sci. Technol.*, **2000**, *34*, 2900-2908.
5. Gingrich, S.E. and Diamond, M.L., *Environ. Sci. Technol.*, **2001**, *35*, 4031-4037.
6. Lam, B., Diamond, M.L., Simpson, A.J., Makar, P.A., Truong, J., and Hernandez-Martinez, N.A., *Atmos. Environ.*, **2005**, *39*, 6578-6586.
7. Simpson, A.J., Lam, B., Diamond, M.L., Donaldson, D.J., Lefebvre, B.A., Moser, A.Q., Williams, A.J., Larin, N.I., and Kvasha, M.P., *Chemosphere*, **2006**, *63*, 142-152.
8. Liu, Q.T., Chen, R., McCarry, B.E., Diamond, M.L., and Bahavar, B., *Environ. Sci. Technol.*, **2003**, *37*, 2340-2349.
9. Liu, Q.T., Diamond, M.L., Gingrich, S.E., Ondov, J.M., Maciejczyk, P., and Stern, G.A., *Environ. Pollution*, **2003**, *122*, 51-61.
10. Bertram, A.K., Ivanov, A.V., Hunter, M., Molina, L.T., and Molina, M.J., *J. Phys. Chem. A*, **2001**, *105*, 9415-9421.
11. Knopf, D.A., Mak, J., Gross, S., and Bertram, A.K., *Geophys. Res. Lett.*, **2006**, *33*.
12. Mak, J., Gross, S., and Bertram, A.K., *Geophys. Res. Lett.*, **2007**, *34*.
13. Vlasenko, A., George, I.J., and Abbatt, J.P.D., *J. Phys. Chem. A*, **2008**, *112*, 1552-1560.
14. Kahan, T.F., Kwamena, N.-O.A., and Donaldson, D.J., *Atmos. Environ.*, **2006**, *40*, 3448-3459.
15. Mmereki, B.T., Chaudhuri, S.R., and Donaldson, D.J., *J. Phys. Chem. A*, **2003**, *107*, 2264-2269.
16. Clifford, D. and Donaldson, D.J., *J. Phys. Chem. A*, **2007**, *111*, 9809-9814.
17. Clifford, D., Donaldson, D.J., Brigante, M., D'Anna, B., and George, C., *Environ. Sci. Technol.*, **2008**, *42*, 1138-1143.
18. Mmereki, B.T. and Donaldson, D.J., *J. Phys. Chem. A*, **2003**, *107*, 11038-11042.

19. Mmereki, B.T., Donaldson, D.J., Gilman, J.B., Eliason, T.L., and Vaida, V., *Atmos. Environ.*, **2004**, *38*, 6091-6103.
20. Kwamena, N.O.A., Earp, M.E., Young, C.J., and Abbatt, J.P.D., *J. Phys. Chem. A*, **2006**, *110*, 3638-3646.
21. Kwamena, N.O.A., Staikova, M.G., Donaldson, D.J., George, I.J., and Abbatt, J.P.D., *J. Phys. Chem. A*, **2007**, *111*, 11050-11058.
22. Kwamena, N.-O.A., Thornton, J.A., and Abbatt, J.P.D., *J. Phys. Chem. A*, **2004**, *108*, 11626-11634.
23. Poschl, U., Letzel, T., Schauer, C., and Niessner, R., *J. Phys. Chem. A*, **2001**, *105*, 4029-4041.
24. Diamond, M.L., Priemer, D.A., and Law, N.L., *Chemosphere*, **2001**, *44*, 1655-1667.
25. Clarke, J.P., *Applying the Multimedia Urban Model to Measured PCBs, PAHs and PBDEs in Urban Areas*. 2006, University of Toronto: Toronto.
26. Mackay, D., *Multimedia Environmental Models: The Fugacity Approach*. 2nd ed. 2001, Boca Raton: Lewis Publishers.
27. Kwamena, N.O.A., Clarke, J.P., Kahan, T.F., Diamond, M.L., and Donaldson, D.J., *Atmos. Environ.*, **2007**, *41*, 37-50.
28. Handley, S.R., Clifford, D., and Donaldson, D.J., *Environ. Sci. Technol.*, **2007**, *41*, 3898-3903.
29. Bartels-Rausch, T. and Donaldson, D.J., *Atmos. Chem. Phys. Disc.*, **2006**. *6* 10713-10731.
30. Chu, L. and Anastasio, C., *J. Phys. Chem. A*, **2003**, *107*, 9594-9602.
31. Vione, D., Maurino, V., Minero, C., Pelizzetti, E., Harrison, M.A.J., Olariu, R.I., and Arsene, C., *Chem. Soc. Rev.*, **2006**, *35*, 441-453.
32. Warneck, P. and Wurzinger, C., *J. Phys. Chem.*, **1988**, *92*, 6278-6283.
33. Zellner, R., Exner, M., and Herrmann, H., *J. Atmos. Chem.*, **1990**, *10*, 411-425.
34. http://www.climate.weatheroffice.ec.gc.ca/climate_normals/ and results_e. html?Province=ALL&StationName=toronto&SearchType=BeginsWith&Lo cateBy=Province&Proximity=25&ProximityFrom=City&StationNumber= &IDType=MSC&CityName=&ParkName=&LatitudeDegrees=&LatitudeM inutes=&LongitudeDegrees=&LongitudeMinutes=&NormalsClass=A&Sel Normals=&StnId=5051&.
35. Mackay, D., Shiu, W.Y., Ma, K.-C., and Lee, S.C., *Physical-Chemical Properties and Environmental Fate for Organic Chemicals on CD-ROM*. 2nd ed. 2006, London: Taylor and Francis.

Chapter 7

Photochemistry of Secondary Organic Aerosol Formed from Oxidation of Monoterpenes

Stephen A. Mang[1], Maggie L. Walser[1,2], Xiang Pan[1], Jia-Hua Xing[1,3], Adam P. Bateman[1], Joelle S. Underwood[1,4], Anthony L. Gomez[1,5], Jiho Park[1,6], and Sergey A. Nizkorodov[1,*]

[1]Department of Chemistry, University of California, Irvine, CA, 92697
[2]Current address: National Council for Science and the Environment, 1707 H Street, Suite 200, Washington, DC 20006
[3]Current address: Kyoto University, Gokasho, Uji, Kyoto, 611–0011 Japan
[4]Current address: Department of Chemistry, Loyola University, 6363 St. Charles Avenue, New Orleans, LA 70118
[5]Current address: Sandia National Laboratories, 7011 East Avenue, Livermore, CA 94551
[6]Current address: Department of Environmental Health, Korea National Open University, Seoul 110–791, Korea

Secondary organic aerosol (SOA) formed from atmospheric oxidation of monoterpenes has a complex and highly dynamic chemical composition. This chapter focuses on photochemical processes occurring inside biogenic SOA particles produced by oxidation of monoterpenes. The SOA material is found to significantly absorb solar radiation in the tropospheric actinic window ($\lambda > 300$ nm). The ensuing photolysis of SOA constituents modifies the SOA chemical composition and leads to emission of small volatile molecules back into the gas-phase. A large number of observed products can be explained by photolysis of organic peroxides and carbonyls. These photochemical processes are likely to occur on atmospherically relevant time scales, and affect chemical properties and toxicity of monoterpene SOA particles.

SOA from Atmospheric Oxidation of Monoterpenes

Biogenic volatile organic compound (BVOC) emissions are an order of magnitude greater than anthropogenic VOC emissions on a global scale (1, 2). Monoterpenes ($C_{10}H_{16}$) comprise a significant portion of BVOC emissions (2, 3), and it is important to understand the atmospheric fates of monoterpenes and their oxidation products. The most commonly occurring monoterpenes (4) are shown in Figure 1. The emission patterns of the various monoterpenes strongly depend on the type of vegetation and on the environmental conditions, however D-limonene makes up the majority of monoterpene emissions over orange groves, while α-pinene and β-pinene dominate over most other kinds of forests, especially those composed of oaks and conifers (3, 5). The involvement of monoterpenes in the formation of tropospheric aerosol was recognized in 1960 by Went (6), who observed that a blue haze formed over pine needles in the presence of ozone. Since then, a great deal has been learned about the reactions between naturally emitted monoterpenes and atmospheric oxidants. It is now well established that oxidation of isoprene and monoterpenes is a significant source of SOA, contributing some 60 Tg C/year to the global SOA budget (7-9).

Figure 2 is a highly simplified diagram of the processes involved in the atmospheric processing of monoterpenes. Gas-phase monoterpenes readily react with the major atmospheric oxidants such as ozone (O_3), hydroxyl radical (OH), and nitrate radical (NO_3). During the day, their concentrations are controlled by OH and O_3, and at night they are controlled by NO_3, with monoterpene lifetimes on the order of a few hours in both cases. Regardless of the initial oxidant, gas-phase oxidation of monoterpenes results in a wide variety of polyfunctional carboxylic acids, ketones, aldehydes, peroxides, and alcohols (10-19). Many of these species have sufficiently low vapor pressure to partition into pre-existing particulate matter. In addition, monoterpenes can partition into aqueous particles or cloud droplets by wet deposition and undergo oxidation via aqueous chemistry (20-23), with droplets subsequently drying out into organic particles.

The chemical composition of monoterpene SOA is generally quite complex, with a number of products still remaining uncharacterized. Pinonaldehyde, pinonic acid, nor-pinonic acid, and pinic acid have been identified as major condensed phase products of the laboratory ozonolysis of α-pinene (24-26). Nopinone was identified as the major particle-phase product of β-pinene ozonolysis (11, 24); norpinic and pinic acids were also produced (11). Limononaldehdye, keto-limonene, keto-limononaldehyde, limononic acid, and keto-limononic acid accounted for ~60% of the observed aerosol mass in ozonolysis of D-limonene (27). The latter three products have also been identified as major constituents of SOA resulting from the photooxidation of D-limonene in the presence of NO_x (28). In the OH-initiated oxidation of D-limonene, both keto-limonene and limononaldehyde were identified as major low volatility products (17). Oxidation of α-pinene and D-limonene by NO_3 yielded pinonaldehyde and limononic acid, respectively, and unidentified nitrates (29).

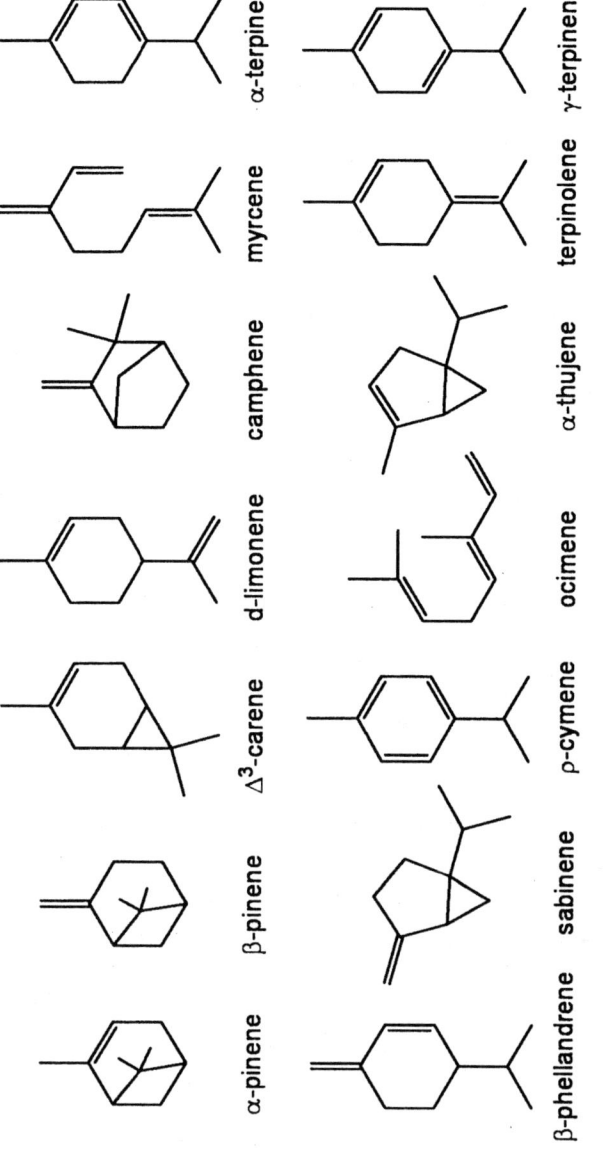

Figure 1. Structures of the most commonly occurring monoterpenes.

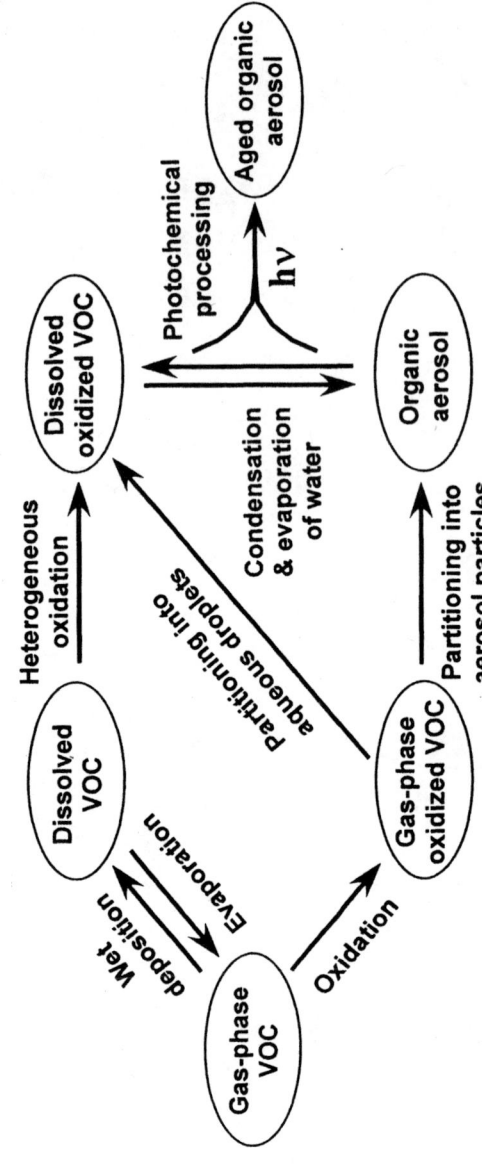

Figure 2. Simplified diagram of atmospheric processing of monoterpenes and their oxidation products.

In addition to the "usual suspects" listed above, several research groups have identified organic peroxides (30-32) and polyfunctional species (12, 14, 15, 33) amongst the products of monoterpene oxidation. There is also increasing evidence that oligomeric species play a key role in the monoterpene SOA formation and subsequent aging (14, 34-43). For example, hundreds of individual compounds in the m/z range of 200-700 were observed by high-resolution mass spectrometry amongst products of ozonolysis of α-pinene (40) and D-limonene (38). Such oligomeric species were attributed to various condensation reactions involving aldehydes, hydroperoxides, carbonyl oxides, and other reactive groups (20, 30, 31, 41, 44).

Photochemical Aging of Organic Aerosol

Aerosol can remain suspended in the atmosphere for several days, especially under dry conditions (9, 45, 46). During this time, particles can significantly change their chemical composition because of exposure to gas-phase oxidants (47) and solar radiation (48, 49). Additionally, reactions between different species inside the particles can modify the composition long after the initial particle growth (36, 37, 50). These "aging" processes may play an important role in determining hygroscopic and toxicological properties of organic particulate matter (8, 47, 51-53).

Most previous work on chemical aging of organic aerosol focused on reactive uptake of gas-phase oxidants (OH, O_3, NO_3, halogen atoms) by representative atmospheric organic materials as summarized in two excellent reviews (47, 54). The main focus of this chapter is on aging caused by photochemical processes occurring *inside* organic particles, which is schematically represented by

$$\text{Monoterpenes} + \text{OH}, O_3, NO_3 \rightarrow \ldots \rightarrow \text{Fresh SOA}$$
$$\text{Fresh SOA} + h\nu \rightarrow \text{Photochemically Aged SOA}$$

Such photochemical aging processes may include direct photodissociation, photoisomerization, photosensitization, and other chemical reactions triggered by absorption of a photon by an SOA constituent. In order for photochemistry to have a significant effect on SOA aging, the following conditions must be satisfied: (1) organic aerosol material must have significant absorption in the tropospheric actinic window ($\lambda > 300$ nm); (2) the yield for photochemical reactions, such as photodissociation, must be large compared to that for fluorescence, vibrational relaxation, geminate recombination, and other non-reactive processes. Our recent experiments (48) suggest that monoterpene SOA satisfy both requirements. Furthermore, direct effect of UV radiation on the yield of monoterpene SOA was observed in several recent studies. During the ozonolysis of α-pinene, exposure to UV radiation decreased the aerosol yield by as much as 20-40% (13, 55), most likely due to a shift in the product volatility distribution to higher volatility species. Presto et al. (13) observed this yield reduction only during particle nucleation, implying that photolysis of non-volatile SOA precursors in the gas phase, prior to nucleation, was responsible. Similar observations were made for D-limonene SOA (56). SOA formed by photochemical oxidation of isoprene was found to be even more susceptible to photodegradation. The fraction of condensable peroxides (36) and the effective yield (57) of SOA particles prepared by photooxidation of isoprene decreased with irradiation time, implying that photochemical aging was occurring after the initial particle formation. The particles lost a significant fraction of their volume as a result of this photochemical aging (57).

Aerosol photochemistry is certainly not limited to direct photolysis of organic species. A review by Vione et al. (58) discussed the importance of secondary photochemical processes in fog and cloud water. Dissolved nitrate, nitrite, iron (III) compounds, and hydrogen peroxide can all undergo photochemical transformations in cloud droplets leading to OH radical and other reactive species, which can then go on to oxidize organics present in the droplet. In this way, even organic molecules that are relatively impervious to direct photolysis can still be chemically modified via photochemistry. For example, photodegradation of several dissolved organic nitrogen species was observed in irradiated fog waters and aerosol aqueous extracts (59, 60). Similar processes may be occurring in monoterpene SOA particles internally mixed with inorganic chromophores, e.g., with nitrates or hydrogen peroxide.

Even if SOA constituents have low initial absorption cross-sections in the tropospheric actinic window, efficient organic absorbers can be generated *in situ* as aerosol ages. For example, uptake of small carbonyls by acidic solutions was shown to lead to the appearance of light-absorbing aldol condensation products (44, 61, 62), and make potentially significant contributions to the absorption index of particles containing sulfuric acid. Our own measurements (49, 63) showed that films of unsaturated organics become photochemically active after exposure to ozone. In summary, there is sufficient evidence to suggest that monoterpene SOA constituents can be photodegraded on atmospherically relevant timescales.

Absorption Spectra of Monoterpene SOA

In principle, one could envision modeling the absorption spectra of SOA as an appropriately weighed linear combination of the spectra of all individual

SOA constituents. This approach is impractical because of the very large number of compounds found in SOA. Alternatively, the absorption spectra of monoterpene SOA can be deduced from spectroscopy of actual particle samples. Measurements of absorption spectra of aerosol are non-trivial because one has to separate the scattering and absorption contributions to the overall extinction (64). Our laboratory uses a simple technique that allows measurements to be made on thin films of SOA material, thus avoiding possible complications from solvents and scattering. The SOA samples are prepared by reaction between monoterpene vapor and ozone in a Teflon chamber filled with dry air. The resulting particles are collected onto a calcium fluoride (CaF_2) window using a single-jet impactor, with the amount of collected material determined gravimetrically. A relatively uniform film is prepared by pressing the collected SOA material between two CaF_2 windows; the films are then analyzed in a UV spectrometer. Spectra are measured relative to clean CaF_2 windows, and corrected for the expected reflection losses.

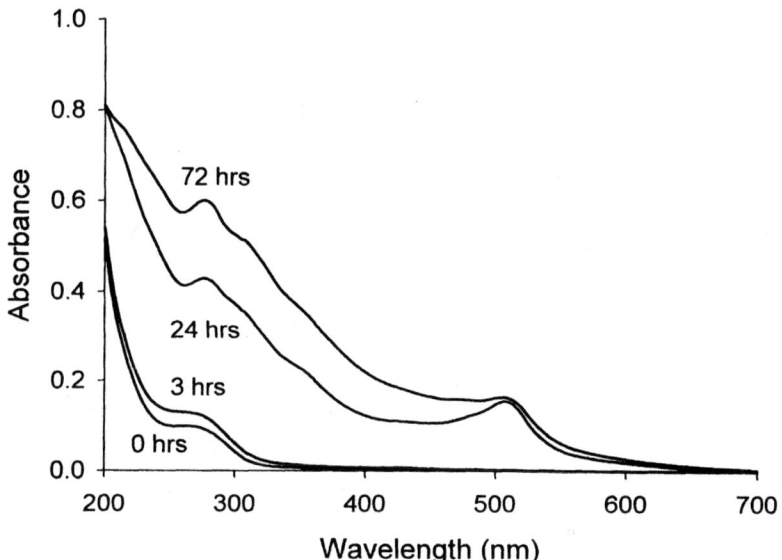

Figure 3. Representative absorption spectrum of organic material formed in ozonolysis of D–limonene as a function of storage time in ambient air in darkness. The changes in the absorption profile are accompanied by a visible transition of the particles from colorless (bottom spectrum) to red-brown.

Figure 3 shows a sample absorption spectrum of a D-limonene SOA sample. Initially the collected SOA material is transparent to visible light, but there is a significant absorption tail extending into the tropospheric actinic range. The shape of the absorption spectrum obtained immediately after collection suggests a mixture of peroxides (sharp decrease in absorption with increasing wavelength) and carbonyl functional groups (weak absorption around 300 nm) in agreement with the standard mechanism of ozonolysis of alkenes. An interesting observation is that the SOA material acquires a reddish color when stored in open air in darkness for several hours. In addition to an unstructured absorption extending over a broad range of wavelengths, two distinct peaks appear at 280 nm and 510 nm. We attribute the short wavelength band to overlapping $n \rightarrow \pi^*$ transitions of all carbonyls in SOA. The long wavelength band has not been assigned. However, the observed time-dependence of the absorption spectrum of SOA strongly suggests that aging processes increase the degree of conjugation in the SOA species. A possible scenario involves cross-aldol condensation reactions of carbonyl compounds (44, 61, 62). Indeed, the aldol condensation of aliphatic aldehydes in concentrated sulfuric acid was shown to produce a large increase in the absorption index in the near UV and visible range (44, 62). Although the SOA material is not as acidic as solutions investigated in Ref. (44, 61, 62), the presence of multiple organic acids in SOA appears to be sufficient for an efficient catalysis of these condensation reactions.

The effective extinction coefficient (ε) of the resulting SOA material can be estimated from the known mass of the collected particles. For a rough estimate, we assume that the SOA film on the CaF_2 window is uniform, that it has a density of 1.5 g/cm^3 (65), and that the average molecular weight of the SOA species is 200 g/mol. This translates into $\varepsilon \sim 10\text{-}100$ L mol^{-1} cm^{-1} for the 510 nm peak after 24 hours of aging. Over the course of a day, the particles transition from being transparent to visible light to being weakly absorbing. Their extinction coefficient for UV radiation also increases by approximately an order of magnitude over this time, giving them a final ε in excess of 100 L mol^{-1} cm^{-1} at 300 nm. This is certainly sufficient for solar radiation-driven photochemistry to occur on atmospherically relevant time scales.

Such an increase in the SOA absorption coefficient resulting from chemical aging has significant implications. Current climate models assign a negative radiative forcing to organic particles because of their light scattering and cloud condensation roles (66, 67). If organic particles can be converted from scatterers to absorbers of visible radiation after spending a few days in the air, their contribution to the radiative forcing could potentially shift from negative to positive. In fact, there is evidence for brown aerosol amplifying warming over Asia (68) but the question of whether biogenic SOA contribute to this effect is still open.

Photodissociation Spectra of Monoterpene SOA

The fact that SOA particles absorb radiation in the near-UV and visible ranges of the solar spectrum does not immediately imply that SOA material is susceptible to solar photodegradation. The absorbed photon energy can be re-emitted as fluorescence or lost to vibrational relaxation. Even if the absorber succeeds in breaking one of its bonds, chances are it will be re-formed because of the likely caging effects in the surrounding SOA matrix. To fully understand the impact of photochemistry on SOA aging, knowledge of photodissociation spectra and photodissociation quantum yields for representative SOA species is required.

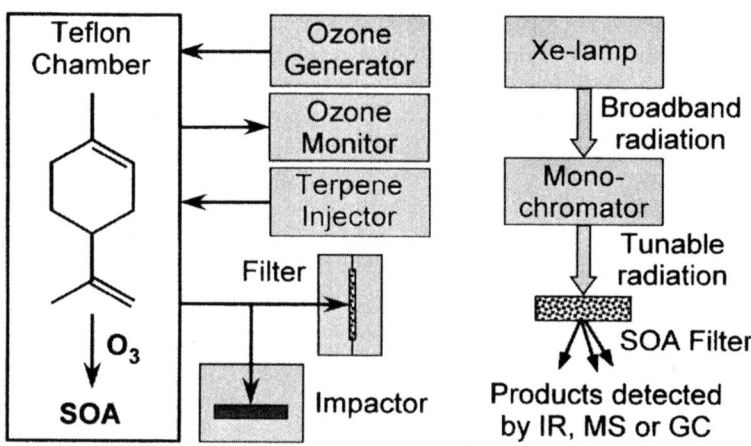

Figure 4. Experimental approach for studying SOA photochemistry. The left section depicts SOA generation, and the right shows the basic principle of photodissociation action spectroscopy for SOA filter samples. IR, MS, and GC refer to infrared cavity ring down spectroscopy, chemical ionization mass spectrometry, and gas chromatography, respectively.

Our laboratory recently developed a new approach for studying photochemistry of organic aerosol materials based on photodissociation action spectroscopy (48, 49, 63). Figure 4 illustrates the basic idea of this approach. SOA particles are synthesized by a dark reaction between a monoterpene and ozone, collected on a quartz filter, and purged with clean nitrogen to remove the most volatile fraction. The filter is then placed in a custom-built photochemical cell where it is exposed to wavelength-tunable UV radiation. The resulting gas-phase photolysis products are sensitively detected in real time either by cavity ring-down infrared spectroscopy (49) or chemical ionization mass-spectrometry.

Alternatively, the products are trapped and analyzed off-line by GC/MS. The relative yield of a specific photolysis product as a function of the excitation wavelength is a *photodissociation action spectrum*. The SOA material can also be extracted from the filter before and after the irradiation and analyzed by electrospray ionization mass-spectrometry in order to assess the degree of chemical change due to photolysis.

For a single isolated molecule, the photodissociation action spectrum represents the product of absorption cross-section with photodissociation quantum yield for a particular product channel. Therefore, comparison between the absorption spectrum and photodissociation spectra recorded for all possible product channels carries rich information about the mechanism of photodissociation. For a complicated mixture of molecules contained in a sample of SOA particles, this comparison is arguably less straightforward. Nevertheless, it can provide useful information about the mechanism of SOA photochemistry, as illustrated in the example given below.

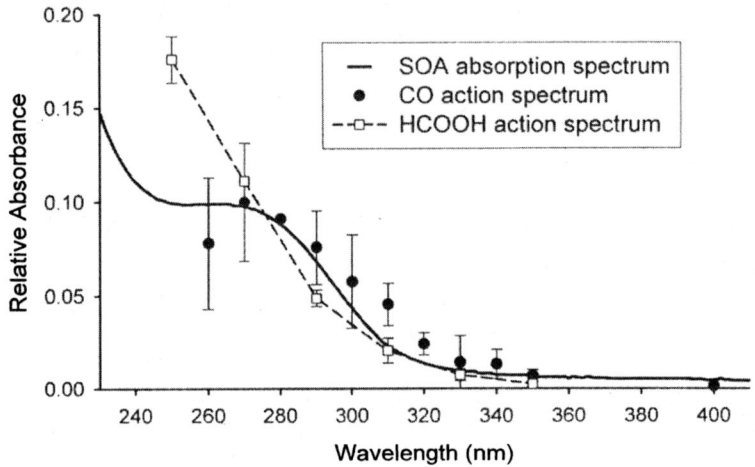

Figure 5. Comparison of the absorption spectrum and two photodissociation action spectra for D-limonene SOA (carbon monoxide and formic acid channels).

Figure 5 shows a comparison between the absorption spectrum of D-limonene SOA, recorded immediately following particle collection to minimize the aging effects, and two photodissociation action spectra. The first action spectrum was recorded by measuring the relative yield of formic acid (HCOOH) as a function of UV irradiation wavelength (48). The second action spectrum tracked the yield of carbon monoxide (CO). Both action spectra were arbitrarily scaled to facilitate the comparison with the absorption spectrum.

The HCOOH and CO action spectra are reproducibly different from each other, reflecting the different photochemical pathways leading to their production (Figure 6). HCOOH is a good marker for photolysis of secondary ozonides formed on ozonolysis of terminal alkenes (48, 49, 63), and therefore the action spectrum recorded for the HCOOH product channel should be similar to the spectra of simple organic peroxides. Peroxides are characterized by a strong n→σ* band in the deep UV that has a broad tail extending into the actinic region. At wavelengths longer than 250 nm, the absorption cross sections of peroxides decay with a dependence that is very nearly exponential. This is consistent with the observed shape of the HCOOH action spectrum. The emission of CO is a signature of carbonyl photolysis (Figure 6). Aliphatic carbonyls are characterized by a weak n→π* band centered in the vicinity of 300 nm. This band is visible in both the absorption spectrum of SOA and the CO action spectrum in agreement with its assignment to the carbonyl functional group.

Photochemistry of carbonyls can potentially generate a large number of additional volatile products via Norrish type I or Norrish type II photochemical cleavage (69); Figure 6 shows just two possible channels. Given the high abundance of carbonyls in monoterpene SOA, one could expect a large number of photoproducts including simple alkanes, alkenes, alcohols, and carbonyls. Indeed, we have been able to detect a large number of products of D-limonene SOA photodegradation by chemical ionization mass-spectrometry and gas-chromatography. A detailed analysis of these data will be reported in a separate publication.

Summary and Future Directions

The most interesting conclusion emerging for this work is that both primary (49, 63) and secondary (48) organic particles can undergo aging processes driven by direct photolysis of organic peroxides, ketones, and aldehydes inside the particles. This photochemical aging has the potential to significantly influence the physicochemical and toxicological properties of organic particles. For example, selective photodegradation of peroxy groups in SOA particles is likely to affect their toxicological properties as peroxides are implicated in many of the adverse health effects of particulate matter (70, 71). The change in composition of organic particles may also affect their ability to act as cloud condensation nuclei, thus indirectly affecting the climate (52, 53). Finally, photochemical reactions in organic particles can generate a radiation-driven flux of small volatile organics from the aerosol phase back into the gas phase (72).

It is clear that we have merely scratched the surface with respect to understanding the mechanism of photochemical aging of monoterpene SOA. The effects of reagent concentration, type of oxidant, humidity, temperature,

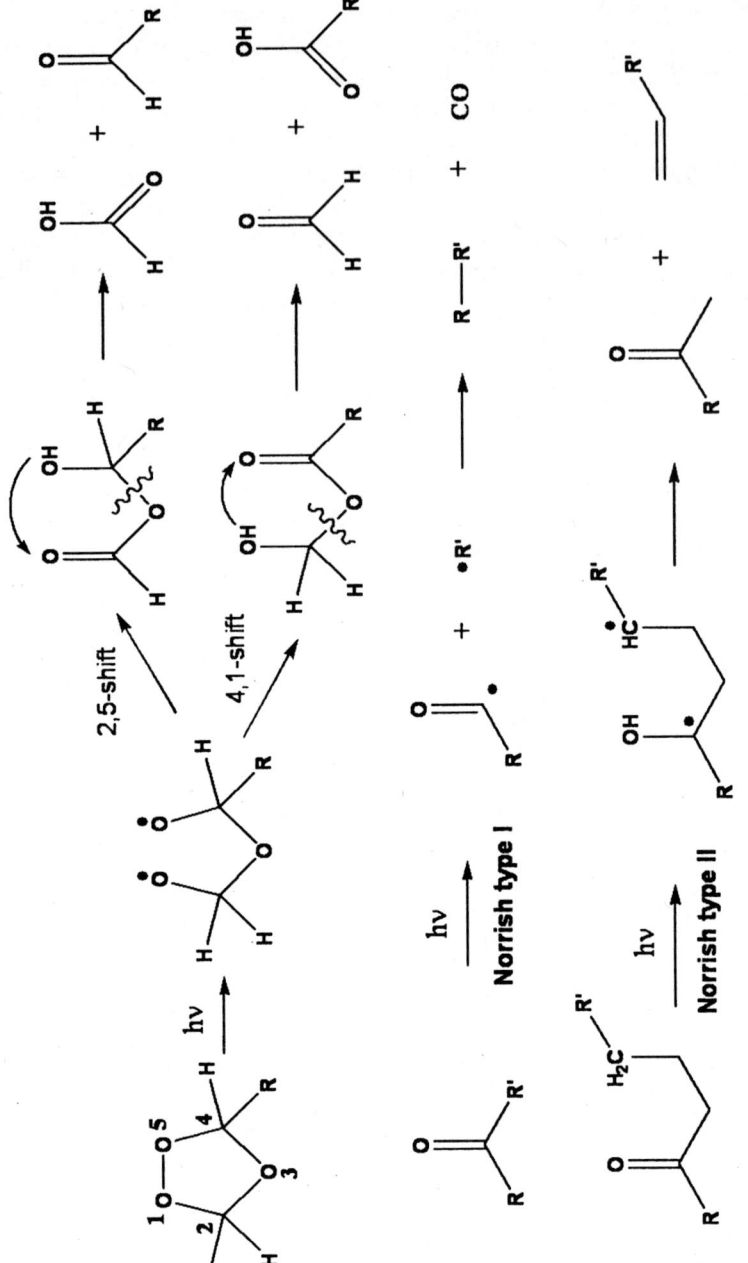

Figure 6. Possible photodissociation pathways for secondary ozonides and carbonyls. For simplicity, only two selected pathways of Norrish type I and II photocleavage are shown. Formic acid and carbon monoxide are markers for photolysis of secondary ozonides and carbonyls, respectively.

and inorganic seed particles used during SOA generation all need to be carefully addressed in future experiments. Similarly, the roles that particle size, chemical heterogeneity, and presence of adsorbed water play in the mechanism of photochemical reactions are currently unknown. We also stress that the role of photochemical processes has been explored only for laboratory generated SOA; the effect of solar radiation on field samples of particulate matter was not studied. On a more fundamental side, the cage effects for photodissociation of simple molecules in characteristic SOA materials, as well as the solvent effects of SOA medium on absorption spectra of simple chromophores will need to be explored in detail. With sufficient quantitative information on photodissociation rates, we anticipate that photochemical aging processes can start to be included in air pollution and regional climate models.

References

1. Lamb, B.; Guenther, A.; Gay, D.; Westberg, H. A National Inventory of Biogenic Hydrocarbon Emissions. *Atm. Environ.* **1987**, *21*, 1695-1705.
2. Guenther, A.; Hewitt, C. N.; Erickson, D.; Fall, R.; Geron, C.; Graedel, T.; Harley, P.; Klinger, L.; Lerdau, M. A Global Model of Natural Volatile Organic Compound Emissions. *J. Geophys. Res. D* **1995**, *100*, 8873-8892.
3. Pio, C.; Valente, A. A. Atmospheric Fluxes and Concentrations of Monoterpenes in Resin-Tapped Pine Forests. *Atm. Environ.* **1998**, *32*, 683-691.
4. Geron, C.; Rasmussen, R.; Arnts, R. R.; Guenther, A. A Review and Synthesis of Monoterpene Speciation from Forests in the United States. *Atm. Environ.* **2000**, *34*, 1761-1781.
5. Christensen, C. S.; Hummelshoj, P.; Jensen, N. O.; Larsen, B.; Lohse, C.; Pilegaard, K.; Skov, H. Determination of the Terpene Flux from Orange Species and Norway Spruce by Relaxed Eddy Accumulation. *Atm. Environ.* **2000**, *34*, 3057-3067.
6. Went, F. W. Blue Hazes in the Atmosphere. *Nature* **1960**, *187*, 641-643.
7. Chung, S. H.; Seinfeld, J. H. Global Distribution and Climate Forcing of Carbonaceous Aerosols. *J. Geophys. Res. D* **2002**, *107*, 4407, doi:10.1029/2001JD001397, 2002
8. Robinson, A. L.; Donahue, N. M.; Shrivastava, M. K.; Weitkamp, E. A.; Sage, A. M.; Grieshop, A. P.; Lane, T. E.; Pierce, J. R.; Pandis, S. N. Rethinking Organic Aerosols: Semivolatile Emissions and Photochemical Aging. *Science* **2007**, *315*, 1259-1262.
9. Goldstein, A. H.; Galbally, I. E. Known and Unexplored Organic Constituents in the Earth's Atmosphere. *Environ. Sci. Technol.* **2007**, *41*, 1514-1521.

10. Kavouras, I. G.; Mihalopoulos, N.; Stephanou, E. G. Formation of Atmospheric Particles from Organic Acids Produced by Forests. *Nature* **1998**, *395*, 683-686.
11. Yu, J.; Cocker, D. R., III; Griffin, R. J.; Flagan, R. C.; Seinfeld, J. H. Gas-Phase Ozone Oxidation of Monoterpenes: Gaseous and Particulate Products. *J. Atm. Chem.* **1999**, *34*, 207-258.
12. Claeys, M.; Szmigielski, R.; Kourtchev, I.; Van der Veken, P.; Vermeylen, R.; Maenhaut, W.; Jaoui, M.; Kleindienst, T. E.; Lewandowski, M.; Offenberg, J. H.; Edney, E. O. Hydroxydicarboxylic Acids: Markers for Secondary Organic Aerosol from the Photooxidation of α-Pinene. *Environ. Sci. Technol.* **2007**, *41*, 1628-1634.
13. Presto, A. A.; Huff Hartz, K. E.; Donahue, N. M. Secondary Organic Aerosol Production from Terpene Ozonolysis. 1. Effect of UV Radiation. *Environ. Sci. Technol.* **2005**, *39*, 7036-7045.
14. Gao, S.; Keywood, M.; Ng, N. L.; Surratt, J.; Varutbangkul, V.; Bahreini, R.; Flagan, R. C.; Seinfeld, J. H. Low-Molecular-Weight and Oligomeric Components in Secondary Organic Aerosol from the Ozonolysis of Cycloalkenes and α-Pinene. *J. Phys. Chem. A* **2004**, *108*, 10147-10164.
15. Dalton, C. N.; Jaoui, M.; Kamens, R. M.; Glish, G. L. Continuous Real-Time Analysis of Products from the Reaction of Some Monoterpenes with Ozone Using Atmospheric Sampling Glow Discharge Ionization Coupled to a Quadrupole Ion Trap Mass Spectrometer. *Anal. Chem.* **2005**, *77*, 3156-3163.
16. Glasius, M.; Duane, M.; Larsen, B. R. Determination of Polar Terpene Oxidation Products in Aerosols by Liquid Chromatography-Ion Trap Mass Spectrometry. *J. Chromatogr. A* **1999**, *833*, 121-135.
17. Hakola, H.; Arey, J.; Aschmann, S. M.; Atkinson, R. Product Formation from the Gas-Phase Reactions of OH Radicals and O_3 with a Series of Monoterpenes. *J. Atm. Chem.* **1994**, *18*, 75-102.
18. Jaoui, M.; Kleindienst, T. E.; Lewandowski, M.; Offenberg, J. H.; Edney, E. O. Identification and Quantification of Aerosol Polar Oxygenated Compounds Bearing Carboxylic or Hydroxyl Groups. 2. Organic Tracer Compounds from Monoterpenes. *Environ. Sci. Technol.* **2005**, *39*, 5661-5673.
19. Leungsakul, S.; Jeffries, H. E.; Kamens, R. M. A Kinetic Mechanism for Predicting Secondary Aerosol Formation from the Reactions of D-Limonene in the Presence of Oxides of Nitrogen and Natural Sunlight. *Atm. Environ.* **2005**, *39*, 7063-7082.
20. Altieri, K. E.; Carlton, A. G.; Lim, H.-J.; Turpin, B. J.; Seitzinger, S. P. Evidence for Oligomer Formation in Clouds: Reactions of Isoprene Oxidation Products. *Environ. Sci. Technol.* **2006**, *40*, 4956-4960.
21. Lim, H.-J.; Carlton, A. G.; Turpin, B. J. Isoprene Forms Secondary Organic Aerosol through Cloud Processing: Model Simulations. *Environ. Sci. Technol.* **2005**, *39*, 4441-4446.

22. Blando, J. D.; Turpin, B. J. Secondary Organic Aerosol Formation in Cloud and Fog Droplets: A Literature Evaluation of Plausibility. *Atm. Environ.* **2000**, *34*, 1623-1632.
23. Jang, M.; Czoschke Nadine, M.; Lee, S.; Kamens Richard, M. Heterogeneous Atmospheric Aerosol Production by Acid-Catalyzed Particle-Phase Reactions. *Science* **2002**, *298*, 814-817.
24. Yokouchi, Y.; Ambe, Y. Aerosols Formed from the Chemical Reaction of Monoterpenes and Ozone. *Atm. Environ.* **1985**, *19*, 1271-1276.
25. Jang, M.; Kamens, R. M. Newly Characterized Products and Composition of Secondary Aerosols from the Reaction of α-Pinene with Ozone. *Atm. Environ.* **1999**, *33*, 459-474.
26. Kamens, R.; Jang, M.; Chien, C.-J.; Leach, K. Aerosol Formation from the Reaction of α-Pinene and Ozone Using a Gas-Phase Kinetics-Aerosol Partitioning Model. *Environ. Sci. Technol.* **1999**, *33*, 1430-1438.
27. Leungsakul, S.; Jaoui, M.; Kamens, R. M. Kinetic Mechanism for Predicting Secondary Organic Aerosol Formation from the Reaction of D-Limonene with Ozone. *Environ. Sci. Technol.* **2005**, *39*, 9583-9594.
28. Jaoui, M.; Corse, E.; Kleindienst, T. E.; Offenberg, J. H.; Lewandowski, M.; Edney, E. O. Analysis of Secondary Organic Aerosol Compounds from the Photooxidation of D-Limonene in the Presence of NO_x and Their Detection in Ambient $PM_{2.5}$. *Environ. Sci. Technol.* **2006**, *40*, 3819-3828.
29. Spittler, M.; Barnes, I.; Bejan, I.; Brockmann, K. J.; Benter, T.; Wirtz, K. Reactions of NO_3 Radicals with Limonene and α-Pinene: Product and SOA Formation. *Atm. Environ.* **2006**, *40*, S116-S127.
30. Tolocka, M. P.; Heaton, K. J.; Dreyfus, M. A.; Wang, S.; Zordan, C. A.; Saul, T. D.; Johnston, M. V. Chemistry of Particle Inception and Growth During α-Pinene Ozonolysis. *Environ. Sci. Technol.* **2006**, *40*, 1843-1848.
31. Docherty, K. S.; Wu, W.; Lim, Y. B.; Ziemann, P. J. Contributions of Organic Peroxides to Secondary Aerosol Formed from Reactions of Monoterpenes with O_3. *Environ. Sci. Technol.* **2005**, *39*, 4049-4059.
32. Norgaard, A. W.; Nojgaard, J. K.; Clausen, P. A.; Wolkoff, P. Secondary Ozonides of Substituted Cyclohexenes: A New Class of Pollutants Characterized by Collision-Induced Dissociation Mass Spectrometry Using Negative Chemical Ionization. *Chemosphere* **2008**, *70*, 2032-2038.
33. Warscheid, B.; Hoffmann, T. Direct Analysis of Highly Oxidized Organic Aerosol Constituents by On-Line Ion Trap Mass Spectrometry in the Negative-Ion Mode. *Rap. Comm. Mass. Spectr.* **2002**, *16*, 496-504.
34. Kalberer, M.; Paulsen, D.; Sax, M.; Steinbacher, M.; Dommen, J.; Prevot, A. S. H.; Fisseha, R.; Weingartner, E.; Frankevich, V.; Zenobi, R.; Baltensperger, U. Identification of Polymers as Major Components of Atmospheric Organic Aerosols. *Science* **2004**, *303*, 1659-1662.
35. Tolocka, M. P.; Jang, M.; Ginter, J. M.; Cox, F. J.; Kamens, R. M.; Johnston, M. V. Formation of Oligomers in Secondary Organic Aerosol. *Environ. Sci. Technol.* **2004**, *38*, 1428-1434.

36. Surratt, J. D.; Murphy, S. M.; Kroll, J. H.; Ng, N. L.; Hildebrandt, L.; Sorooshian, A.; Szmigielski, R.; Vermeylen, R.; Maenhaut, W.; Claeys, M.; Flagan, R. C.; Seinfeld, J. H. Chemical Composition of Secondary Organic Aerosol Formed from the Photooxidation of Isoprene. *J. Phys. Chem. A* **2006**, *110*, 9665-9690.
37. Baltensperger, U.; Kalberer, M.; Dommen, J.; Paulsen, D.; Alfarra, M. R.; Coe, H.; Fisseha, R.; Gascho, A.; Gysel, M.; Nyeki, S.; Sax, M.; Steinbacher, M.; Prevot, A. S. H.; Sjogren, S.; Weingartner, E.; Zenobi, R. Secondary Organic Aerosols from Anthropogenic and Biogenic Precursors. *Faraday Discuss.* **2005**, *130*, 265-278.
38. Walser, M. L.; Dessiaterik, Y.; Laskin, J.; Laskin, A.; Nizkorodov, S. A. High-Resolution Mass Spectrometric Analysis of Secondary Organic Aerosol Produced by Ozonation of Limonene. *Phys. Chem. Chem. Phys.* **2008**, *10*, 1009-1022.
39. Gao, S.; Ng, N. L.; Keywood, M.; Varutbangkul, V.; Bahreini, R.; Nenes, A.; He, J.; Yoo, K. Y.; Beauchamp, J. L.; Hodyss, R. P.; Flagan, R. C.; Seinfeld, J. H. Particle Phase Acidity and Oligomer Formation in Secondary Organic Aerosol. *Environ. Sci. Technol.* **2004**, *38*, 6582-6589.
40. Reinhardt, A.; Emmenegger, C.; Gerrits, B.; Panse, C.; Dommen, J.; Baltensperger, U.; Zenobi, R.; Kalberer, M. Ultrahigh Mass Resolution and Accurate Mass Measurements as a Tool to Characterize Oligomers in Secondary Organic Aerosols. *Anal. Chem.* **2007**, *79*, 4074-4082.
41. Heaton, K. J.; Dreyfus, M. A.; Wang, S.; Johnston, M. V. Oligomers in the Early Stage of Biogenic Secondary Organic Aerosol Formation and Growth. *Environ. Sci. Technol.* **2007**, *41*, 6129-6136.
42. Petters, M. D.; Kreidenweis, S. M.; Snider, J. R.; Koehler, K. A.; Wang, Q.; Prenni, A. J.; Demott, P. J. Cloud Droplet Activation of Polymerized Organic Aerosol. *Tellus B: Chem. Phys. Meteorol.* **2006**, *58B*, 196-205.
43. Hearn, J. D.; Smith, G. D. Reactions and Mass Spectra of Complex Particles Using Aerosol CIMS. *Intern. J. Mass Spectr.* **2006**, *258*, 95-103.
44. Noziere, B.; Esteve, W. Light-Absorbing Aldol Condensation Products in Acidic Aerosols: Spectra, Kinetics, and Contribution to the Absorption Index. *Atmos. Environ.* **2007**, *41*, 1150-1163.
45. Robinson, A. L.; Subramanian, R.; Donahue, N. M.; Bernardo-Bricker, A.; Rogge, W. F. Source Apportionment of Molecular Markers and Organic Aerosol - 1. Polycyclic Aromatic Hydrocarbons and Methodology for Data Visualization. *Environ. Sci. Technol.* **2006**, *40*, 7803-7810.
46. Gaffney, J. S.; Marley, N. A.; Cunningham, M. M. Natural Radionuclides in Fine Aerosols in the Pittsburgh Area. *Atm. Environ.* **2004**, *38*, 3191-3200.
47. Rudich, Y.; Donahue, N. M.; Mentel, T. F. Aging of Organic Aerosol: Bridging the Gap between Laboratory and Field Studies. *Annu. Rev. Phys. Chem.* **2007**, *58*, 321-352.

48. Walser, M. L.; Park, J.; Gomez, A. L.; Russell, A. R.; Nizkorodov, S. A. Photochemical Aging of Secondary Organic Aerosol Particles Generated from the Oxidation of D-Limonene. *J. Phys. Chem. A* **2007**, *111*, 1907-1913.
49. Gomez, A.; Park, J.; Walser, M.; Lin, A.; Nizkorodov, S. A. UV Photodissociation Spectroscopy of Oxidized Undecylenic Acid Films. *J. Phys. Chem. A* **2006**, *110*, 3584-3592.
50. Sax, M.; Zenobi, R.; Baltensperger, U.; Kalberer, M. Time Resolved Infrared Spectroscopic Analysis of Aerosol Formed by Photo-Oxidation of 1,3,5-Trimethylbenzene and α-Pinene. *Aerosol Sci. Technol.* **2005**, *39*, 822-830.
51. Petters, M. D.; Prenni, A. J.; Kreidenweis, S. M.; DeMott, P. J.; Matsunaga, A.; Lim, Y. B.; Ziemann, P. J. Chemical Aging and the Hydrophobic-to-Hydrophilic Conversion of Carbonaceous Aerosol. *Geophys. Res. Lett.* **2006**, *33*, L24806, doi:10.1029/2006GL027249, 2006.
52. Kanakidou, M.; Seinfeld, J. H.; Pandis, S. N.; Barnes, I.; Dentener, F. J.; Facchini, M. C.; Van Dingenen, R.; Ervens, B.; Nenes, A.; Nielsen, C. J.; Swietlicki, E.; Putaud, J. P.; Balkanski, Y.; Fuzzi, S.; Horth, J.; Moortgat, G. K.; Winterhalter, R.; Myhre, C. E. L.; Tsigaridis, K.; Vignati, E.; Stephanou, E. G.; Wilson, J. Organic Aerosol and Global Climate Modelling: A Review. *Atm. Chem. Phys.* **2005**, *5*, 1053-1123.
53. Sun, J.; Ariya, P. A. Atmospheric Organic and Bio-Aerosols as Cloud Condensation Nuclei (CCN): A Review. *Atm. Environ.* **2006**, *40*, 795-820.
54. Rudich, Y. Laboratory Perspectives on the Chemical Transformations of Organic Matter in Atmospheric Particles. *Chem. Rev.* **2003**, *103*, 5097-5124.
55. Presto, A. A.; Donahue, N. M. Investigation of α-Pinene + Ozone Secondary Organic Aerosol Formation at Low Total Aerosol Mass. *Environ. Sci. Technol.* **2006**, *40*, 3536-3543.
56. Zhang, J.; Huff Hartz, K. E.; Pandis, S. N.; Donahue, N. M. Secondary Organic Aerosol Formation from Limonene Ozonolysis: Homogeneous and Heterogeneous Influences as a Function of NO_x *J. Phys. Chem. A* **2006**, *110*, 11053-11063.
57. Kroll, J. H.; Ng, N. L.; Murphy, S. M.; Flagan, R. C.; Seinfeld, J. H. Secondary Organic Aerosol Formation from Isoprene Photooxidation. *Environ. Sci. Technol.* **2006**, *40*, 1869-1877.
58. Vione, D.; Maurino, V.; Minero, C.; Pelizzetti, E.; Harrison, M. A. J.; Olariu, R.-I.; Arsene, C. Photochemical Reactions in the Tropospheric Aqueous Phase and on Particulate Matter. *Chem. Soc. Rev.* **2006**, *35*, 441-453.
59. Anastasio, C.; McGregor, K. G. Photodestruction of Dissolved Organic Nitrogen Species in Fog Waters. *Aerosol Sci. Technol.* **2000**, *32*, 106-119.

60. Zhang, Q.; Anastasio, C. Conversion of Fogwater and Aerosol Organic Nitrogen to Ammonium, Nitrate, and NO_x During Exposure to Simulated Sunlight and Ozone. *Environ. Sci. Technol.* **2003**, *37*, 3522-3530.
61. Noziere, B.; Esteve, W. Organic Reactions Increasing the Absorption Index of Atmospheric Sulfuric Acid Aerosols. *Geophys. Res. Lett.* **2005**, *32*, L03812, doi:10.1029/2004GL021942, 2005.
62. Casale, M. T.; Richman, A. R.; Elrod, M. J.; Garland, R. M.; Beaver, M. R.; Tolbert, M. A. Kinetics of Acid-Catalyzed Aldol Condensation Reactions of Aliphatic Aldehydes. *Atm. Environ.* **2007**, *41*, 6212-6224.
63. Park, J.; Gomez, A. L.; Walser, M. L.; Lin, A.; Nizkorodov, S. A. Ozonolysis and Photolysis of Alkene-Terminated Self-Assembled Monolayers on Quartz Nanoparticles: Implications for Photochemical Aging of Organic Aerosol Particles. *Phys. Chem. Chem. Phys.* **2006**, *8*, 2506-2512.
64. Sheridan, P.; Arnott, W.; Ogren, J.; Andrews, E.; Atkinson, D.; Covert, D.; Moosmueller, H.; Petzold, A.; Schmid, B.; Strawa, A.; Varma, R.; Virkkula, A. The Reno Aerosol Optics Study: An Evaluation of Aerosol Absorption Measurement Methods. *Aerosol Sci. Technol.* **2005**, *39*, 1-16.
65. Kostenidou, E.; Pathak, R. K.; Pandis, S. N. An Algorithm for the Calculation of Secondary Organic Aerosol Density Combining AMS and SMPS Data. *Aerosol Sci. Technol.* **2007**, *41*, 1002-1010.
66. Penner, J. E.; Andreae, M. O.; Annegarn, H.; Barrie, L.; Feichter, J.; Hegg, D. A.; Jayaraman, A.; Leaitch, R.; Murphy, D. M.; Nganga, J.; Pitari, G., Aerosols, Their Direct and Indirect Effects. In *Climate Change 2001: The Scientific Basis. Contribution of Working Group I to the Third Assessment Report of the Intergovernmental Panel on Climate Change.*, Houghton, J. T.; Ding, Y.; Griggs, D. J.; Noguer, M.; van der Linden, P. J.; Dai, X.; Maskell, K.; Johnson, C. A., Eds. Cambridge University Press: New York, USA, 2001.
67. Solomon, S.; Qin, D.; Manning, M.; Chen, Z.; Marquis, M.; Averyt, K. B.; Tignor, M.; Miller, H. L. *Climate Change 2007: The Physical Science Basis. Contribution of Working Group I to the Fourth Assessment Report of the Intergovernmental Panel on Climate Change.* Cambridge University Press: New York, USA, 2007.
68. Ramanathan, V.; Ramana, M. V.; Roberts, G.; Kim, D.; Corrigan, C.; Chung, C.; Winker, D. Warming Trends in Asia Amplified by Brown Cloud Solar Absorption. *Nature* **2007**, *448*, 575-578.
69. *Photochemistry*; Calvert, J. G.; Pitts, J. N. John Wiley & Sons, Inc., New York, 1966; 899 pages.
70. Friedlander, S. K.; Yeh, E. K. The Submicron Atmospheric Aerosol as a Carrier of Reactive Chemical Species: Case of Peroxides. *Appl. Occup. Environ. Hygiene* **1998**, *13*, 416-420.
71. Weschler, C. J. Ozone's Impact on Public Health: Contributions from Indoor Exposures to Ozone and Products of Ozone-Initiated Chemistry. *Env. Health Persp.* **2006**, *114*, 1489-1496.

72. Kwan, A. J.; Crounse, J. D.; Clarke, A. D.; Shinozuka, Y.; Anderson, B. E.; Crawford, J. H.; Avery, M. A.; McNaughton, C. S.; Brune, W. H.; Singh, H. B.; Wennberg, P. O. On the Flux of Oxygenated Volatile Organic Compounds from Organic Aerosol Oxidation. *Geophys. Res. Lett.* **2006**, *33*, L15815, doi:10.1029/2006GL026144, 2006.

Chapter 8

Effect of Highly Concentrated Dry $(NH_4)_2SO_4$ Seed Aerosols on Ozone and Secondary Organic Aerosol Formation in Aromatic Hydrocarbon/NO_x Photooxidation Systems

Zifeng Lu[1], Kiming Hao[1,*], Junhua Li[1], and Hideto Takekawa[2]

[1]Department of Environmental Science and Engineering,
Tsinghua University, Beijing 100084, China
[2]Engine Combustion and Environmental Laboratory, Mechanical Engineering Department, Toyota Central Research and Development Laboratory, Nagakute, Aichi 480–1192, Japan

Ozone and secondary organic aerosol (SOA) formation from toluene, m-xylene and 1,2,4-trimethylbenzene/NO_x photo-oxidation are studied in a 2 m^3 temperature-controlled smog chamber in the presence of highly concentrated (>20 μm^3 cm^{-3}) dry $(NH_4)_2SO_4$ seed aerosols. The results indicate that the presence of highly concentrated dry $(NH_4)_2SO_4$ aerosols has neither observable effect on ozone formation nor gas-phase reactions, but it does enhance SOA generation and increase SOA yield, which is found to be positively related with the $(NH_4)_2SO_4$ surface concentration. It is proposed that the heterogeneous reactions occurring at the $(NH_4)_2SO_4$ particle surface can cause the incondensable compounds (ICs) to oligomerize to condensable compounds (CCs), which could explain the dry $(NH_4)_2SO_4$ seeds effect on SOA formation.

Aromatic hydrocarbons, which reflect anthropogenic activities, are an important class of volatile organic compounds (VOCs) in the atmosphere (*1, 2*). In urban areas, aromatic hydrocarbons are the second largest contributor (next to alkanes) to total non-methane VOCs, and the largest contributor to maximum incremental reactivity (MIR), which approximates the ozone formation potential (*2*). The chemical oxidation of aromatic hydrocarbons through atmospheric reactions can also lead to the formation of secondary organic aerosols (SOAs), which are a major contributor to fine particulate matter ($PM_{2.5}$, PM with aerodynamic diameter less than 2.5 μm) (*3, 4*), and raises several environmental and geophysical concerns (*5, 6*).

Using smog chambers, a number of laboratory experiments have been conducted to study SOA formation in the past two decades, and most of these experiments were conducted with inorganic seed aerosols to facilitate initial condensation (*7-12*). Cocker et al. (*7, 8*) investigated the effect of water on SOA gas-particle partitioning in both α-pinene/ozone and aromatic photooxidation systems, and found that the measured SOA yield was not affected by the presence of dry inorganic seed aerosols, even at elevated relative humidity (RH). Recent studies (*13, 14*) showed that seed particle acidity enhances SOA yield by accelerating the formation of larger oligomers, and a composite seed aerosol of $(NH_4)_2SO_4$ and H_2SO_4 was usually used as an acid catalyst to catalyze the heterogeneous reactions of carbonyl compounds (*15, 16*).

However, most of these previous studies were conducted with relatively low concentrations of inorganic seed aerosols. The initial concentration of inorganic seed particles was typically 1000-20000 particles cm^{-3} with a number mean diameter of approximately 50-100 nm (*7-12*). Assuming the initial seed aerosol had a log-normal size distribution with a geometric standard deviation (σ_g) of 1.6, the volume concentration of inorganic seed aerosols in these studies was no more than 20 μm^3 cm^{-3}. However, highly concentrated aerosols always exist in the ambient atmosphere of developing cities. For example, Beijing, the capital of China, is experiencing serious PM pollution. The annual average concentration of $PM_{2.5}$ in Beijing was about 100 μg m^{-3}, and during smog episode days it would exceed 300 μg m^{-3}, in which ionic species account for one third (*17, 18*).

A new 2 m^3 temperature controlled smog chamber system was constructed in Tsinghua University to study photochemical reactions under high PM contaminated condition specified for Beijing. In this work, the effect of highly concentrated (>20 μm^3 cm^{-3}) dry $(NH_4)_2SO_4$ seed aerosols on ozone and SOA formation in aromatic hydrocarbon/NO_x photooxidation systems are investigated. $(NH_4)_2SO_4$ is chosen as the seed aerosol because it is a major (sometimes the most abundant) ionic species in Beijing during dust and haze days (*18*). Toluene, *m*-xylene and 1,2,4-trimethylbenzene are selected as the surrogate of aromatic hydrocarbons because they are the three most abundant aromatics in the urban air (*2*) and have been widely studied previously (*8-12*).

Experimental Section

The experiments were performed in a smog chamber which is described in detail elsewhere (*19*). The schematic of the chamber system is shown in Figure 1. The cuboid reactor, which has a volume of 2 m^3 and a surface-to-volume ratio of 5 m^{-1}, is constructed with 50 μm-thick FEP-Teflon film. The reactor is situated in a temperature controlled room (Escpec SEWT-Z-120) where the temperature can be controlled in the range of 10 to 60 °C with an accuracy of ± 0.5 °C. The reactor is irradiated by 40 blacklights (GE F40T12/BLB), and the NO$_2$ photolysis rate was calculated to be 0.21 min^{-1}.

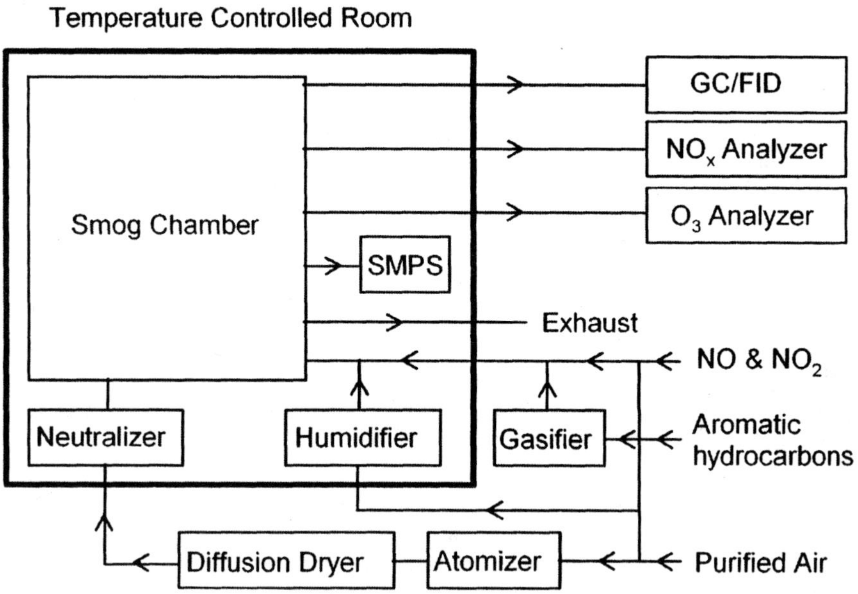

Figure 1. Schematic of the smog chamber system.

Prior to each experiment, the chamber was flushed for 40 hours with purified air at a flow rate of 15 L min^{-1}. Characterization experiments were done to estimate the reactivity of the purified air and no detectable effect on the aerosol formation was found (*19*). In the last several hours' flushing, humid air was introduced into the chamber at a flow rate of 10 L min^{-1} to obtain specific RH. Using a constant output atomizer (TSI Model 3076), seed particles were generated by atomizing an aqueous (NH$_4$)$_2$SO$_4$ solution, and subsequently passed through a diffusion dryer (TSI Model 3062) and a neutralizer (TSI Model 3077) to decrease the humidity and the particle charge. The output aerosol stream, with a RH of ~25 %, was then injected into the chamber, where RH was about 60%. Since the efflorescence relative humidity (ERH) and the deliquescence relative humidity (DRH) of $(NH_4)_2SO_4$ at 303 K are about 35% and 80%, respectively, the introduced $(NH_4)_2SO_4$ seed particles were considered crystalline. To obtain highly concentrated seed particles, the typical initial concentrations of the inorganic particles were 20000-35000 particles cm^{-3} with a number mean diameter of 100-150 nm. Aromatic hydrocarbons, NO and NO_2 were then injected into the chamber. The experiment was then initiated by turning on the blacklights.

The concentration of aromatic hydrocarbon was measured every 15 minutes using gas chromatograph (GC, Beifen SP-3420) equipped with a DM-5 column (30m×0.53mm×1.5μm, Dikma) and a flame ionization detector (FID). NO_x analyzer (Thermo Environmental Instruments, Model 42C) and O_3 analyzer (Thermo Environmental Instruments, Model 49C) were used to monitor NO_x and O_3 with intervals of 1 minute and 15 minutes, respectively. A scanning mobility particle sizer (SMPS, TSI 3936) was used to measure the size distribution and number concentration of aerosol in the range of 17 to 1000 nm with a 6-min cycle. The volume concentration of aerosol is obtained by assuming the particles are geometrically spherical.

Results and Discussions

Calculation of SOA Yields

SOA Yield and Empirical SOA Formation Model

The fractional aerosol yield (Y) is widely used as an empirical expression to represent the aerosol formation potential of an individual reactive organic gas (ROG) (*20-22*), and is defined as the ratio of generated organic aerosol concentration (M_o, μg m^{-3}) to the reacted ROG concentration (ΔROG, μg m^{-3}). Odum et al. (*10, 11*) found that Y largely depends on the amount of organic aerosol mass present, and explained this behavior by using gas/particle absorptive partitioning theory of organic compounds (*23, 24*). In this framework, the gas/particle partitioning process of organic products from oxidation of ROGs is dominated by absorption, which can occur even when the organic products in gas phase are undersaturated. The yield is expressed by the following equation:

$$Y = M_o \sum_i \frac{\alpha_i K_{om,i}}{1 + K_{om,i} M_o} \quad (1)$$

where α_i is a mass-based stoichiometric coefficient for the reaction product i and $K_{om,i}$ (m^3 µg^{-1}) is a normalized partitioning coefficient of product i. Eq 1 can be simplified to one-product model (i.e., i=1) or two-product model (i.e., i=2) by assuming that all semi-volatile products can be classified into one or two groups by their condensabilities. Parameters (α and K_{om}) are obtained by fitting experimentally determined SOA yields with least square method. Since hundreds of products are generated from ROGs oxidation, parameters obtained numerically have no actual physical meaning, but represent the overall properties of all the products (*10*).

Particle Wall Loss Correction

The SOA yields were calculated at the time when measured particle volume concentration reached maximum. A density of 1 g cm^{-3} was assumed to convert organic aerosol volume concentration to mass concentration. Here, the measured aerosol concentration should be corrected to take into account the particle wall loss due to deposition on the Teflon film. The correction method described in Takekawa et al. (*12*) is adopted since most experiments in this work were conducted with highly concentrated preexisting seed aerosols and coagulation may not be neglected. In this method, the aerosol deposition rate constant (k_{dep}, h^{-1}) is related with particle diameter (d_p, nm) by a four-parameter equation:

$$k_{dep}(d_p) = ad_p^b + c/d_p^d \quad (2)$$

k_{dep} values of different d_p are determined by monitoring the particle number decay under dark condition at a low number concentration (< 1000 particles cm^{-3}) to avoid coagulation. The optimized values of parameter a~d are 4.37×10^{-4}, 0.925, 84.3 and 1.40, respectively. It should be noticed that the deposited aerosol concentration estimated by this method might introduce some error due to the low particle number concentration (*12*).

Survey Experiments for Dry (NH$_4$)$_2$SO$_4$ Seed Effect

Six survey experiments (S1~S6) were first conducted to test whether the highly concentrated dry (NH$_4$)$_2$SO$_4$ seed aerosols have effects on ozone and SOA formation in three aromatic hydrocarbon/NO$_x$ photooxidation systems, and their initial conditions and results are summarized in Table I. For each hydrocarbon, two similar experiments were conducted under identical initial conditions except the presence of (NH$_4$)$_2$SO$_4$ aerosols. Figure 2 shows the variations of NO_x-NO, O_3 and $PM_{corrected}$-PM_0 (i.e. the generated SOA) concentration with time in these experiments. The results indicate that:

- The presence of highly concentrated dry $(NH_4)_2SO_4$ aerosols has no significant effect on gas-phase reactions. The trend of NO_x-NO and O_3 concentration variations with time is nearly the same for the particle-free and particle-introduced experiments, and the same statement can be made for NO and hydrocarbon consumptions. The small difference of the O_3 and NO_x-NO profiles may be due to the small discrepancy of initial experimental conditions.
- The presence of highly concentrated dry $(NH_4)_2SO_4$ aerosols obviously enhances organic aerosol generation and increases SOA yield. The surface of dry $(NH_4)_2SO_4$ particles may not be inert, and may influence the gas-particle partitioning process.

Table I. Initial Conditions and Results of Survey Experiments (303K)

Hydrocarbons		Toluene		m-Xylene		1,2,4-TMB[a]	
Experiment No.		S1	S2	S3	S4	S5	S6
$[NO]_0$	(ppb)	151	151	161	166	136	137
$[NO_x]_0$	(ppb)	302	303	333	348	277	282
$[HC]_0$	(ppm)	1.83	1.90	2.00	2.07	1.97	2.14
$[PM]_0$	($\mu m^3 \cdot cm^{-3}$)	0	95	0	74	0	51
$[HC]_0/[NO_x]_0$	(ppm ppm^{-1})	6.1	6.3	6.0	6.0	7.1	7.6
RH	(%)	61	61	53	51	57	59
$[O_3]_{max}$	(ppb)	270	263	412	410	501	498
Time$_{[O_3]max}$	(h)	3.5	3.5	1.5	1.5	2.8	2.8
$[NO_x\text{-}NO]_{max}$	(ppb)	230	232	286	297	230	232
Time$_{[NO_x\text{-}NO]max}$	(h)	1.4	1.4	0.5	0.6	1.0	1.0
$\Delta[HC]_{reacted}$	(ppm)	0.34	0.32	0.45	0.47	0.44	0.43
ΔM_0	($\mu g \cdot m^{-3}$)	88	122	148	249	44	69
SOA Yield	(%)	7.1	10.6	7.6	12.3	2.1	3.3

[a] 1,2,4-Trimethylbenzene.

Effect of Dry $(NH_4)_2SO_4$ Seeds on SOA Formation

Comparison of the SOA Yields between Aromatic Hydrocarbons

Using SOA yield curves calculated from eq 1, a series of experiments were further conducted to investigate how dry $(NH_4)_2SO_4$ particles affect SOA formation. To avoid the potential impacts of different parameters on SOA formation, the hydrocarbon to NO_x ratio, temperature and RH were fixed. The detailed experimental conditions are described later. Experiments in the absence of $(NH_4)_2SO_4$ seeds were first performed to obtain a base SOA yield curve for each hydrocarbon. Figure 3 shows the particle-free aerosol yield curves for three aromatic hydrocarbons. The yield curves are produced by fitting the experimental data to a one-product model based on eq 1. Values of α and K_{om} for each hydrocarbon are as follows: 0.148 and 0.022 for toluene, 0.109 and 0.017 for m-xylene, and 0.062 and 0.011 for 1,2,4-trimethylbenzene.

Figure 3 shows that among the three aromatic hydrocarbons, the SOA yield of toluene is the highest, and that of 1,2,4-trimethylbenzene is the lowest, with m-xylene in the middle. This result is consistent with previous work of Odum et al. (*11*) and Takekawa et al. (*12*). Odum et al. (*11*) indicated that the aromatic species containing one or fewer methyl substituent have higher SOA yield than those containing two or more methyl substituents. Thus, toluene is classified as a "high-yield" aromatic species, and m-xylene and 1,2,4-trimethylbenzene as "low-yield" species. Takekawa et al. (*12*) pointed out that an aromatic hydrocarbon with a lower reaction rate constant with OH radicals may have a higher secondary to primary reaction rate ratio, which is proportional to the SOA yields. Since the order of reaction rate constants of the three aromatic hydrocarbons with OH radicals are toluene, m-xylene and 1,2,4-trimethylbenzene from low to high, our experimental results are in good agreement with Takekawa's theory.

Toluene/NO_x Photooxidation Systems

The initial experimental conditions and results for toluene experiments are shown in Table II, and the SOA yields versus M_o are shown in Figure 4. The aerosol yields obtained with dry $(NH_4)_2SO_4$ aerosols are higher than particle-free ones, and it confirms the conclusion derived from survey experiments that the presence of highly concentrated dry $(NH_4)_2SO_4$ aerosols increases SOA yields.

As shown in Figure 4, the presence of dry $(NH_4)_2SO_4$ seed aerosols increases SOA yields, and the data appears to fall into several different yield curves. To identify which factor is the intrinsic cause for the effect of dry $(NH_4)_2SO_4$, detailed seed information of experiments AS1~AS9 are listed in Table III. $PM_{v,g}$ and $PM_{s,g}$ in Table III stand for the volume and surface PM (i.e. $(NH_4)_2SO_4$) concentrations when SOA starts to generate. Here, the time when SOA begins to form is selected at the time when measured PM decreases to a minimum. Before SOA generation, the measured aerosol volume concentration will continue decreasing due to the deposition. Since the SOA generation process is very fast and the measurement interval of SMPS is 6 minutes, the PM minimum point is approximately equal to the real SOA formation point. After checking the data, it is reasonable to categorize these data into four groups based

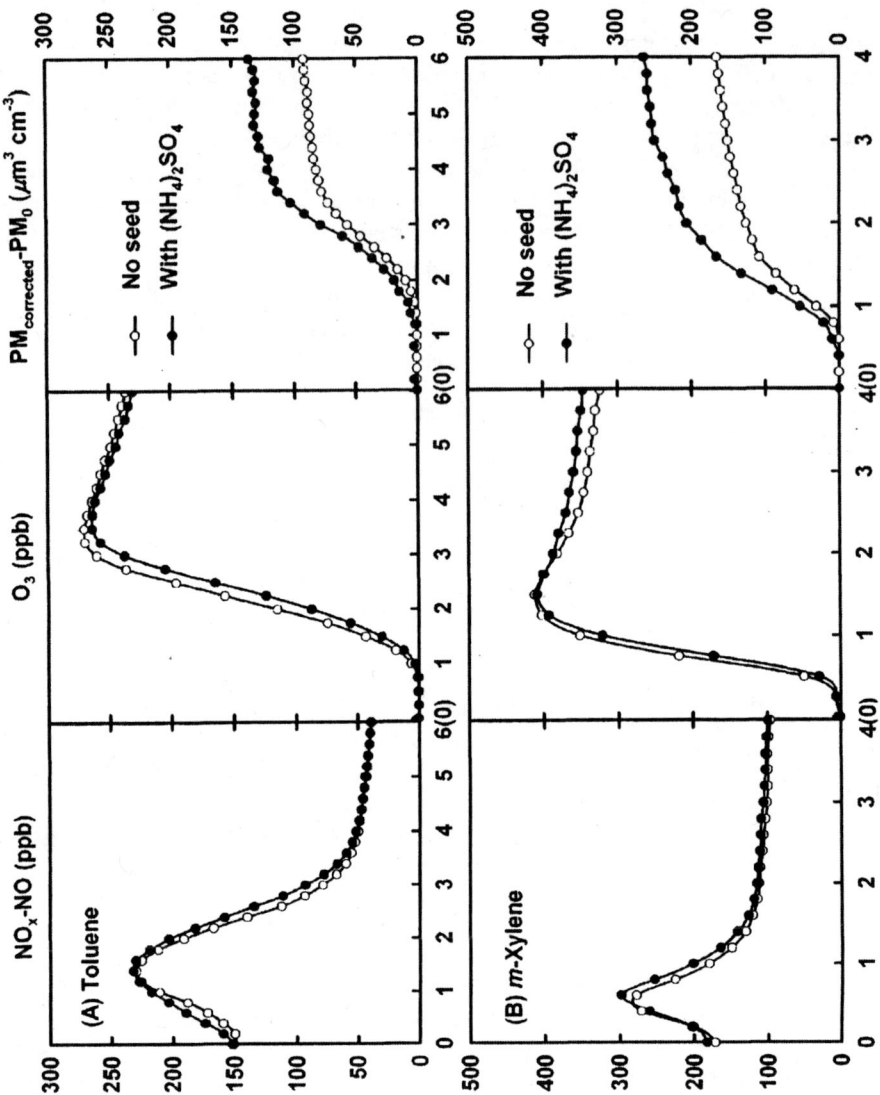

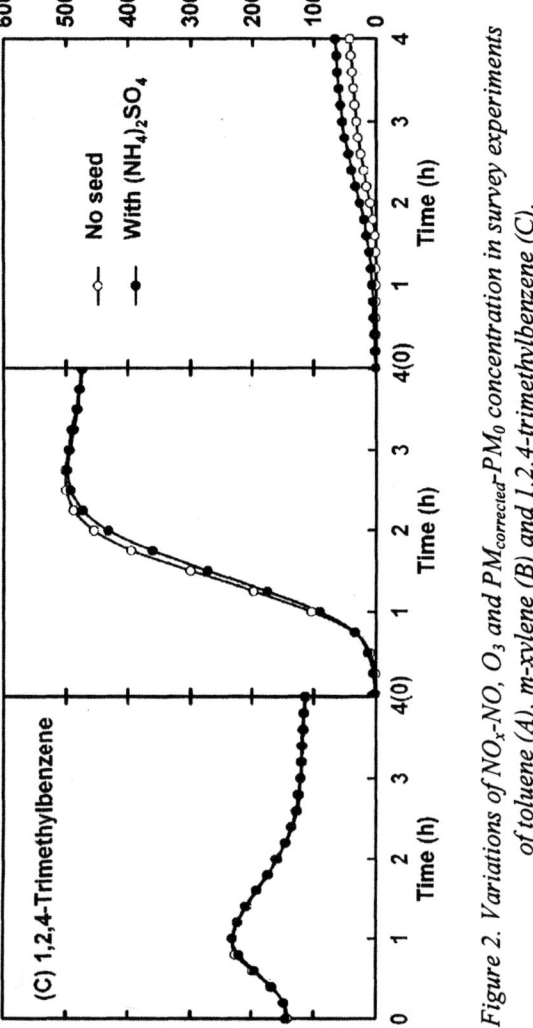

Figure 2. Variations of NO_x-NO, O_3 and $PM_{corrected}$-PM_0 concentration in survey experiments of toluene (A), m-xylene (B) and 1,2,4-trimethylbenzene (C).

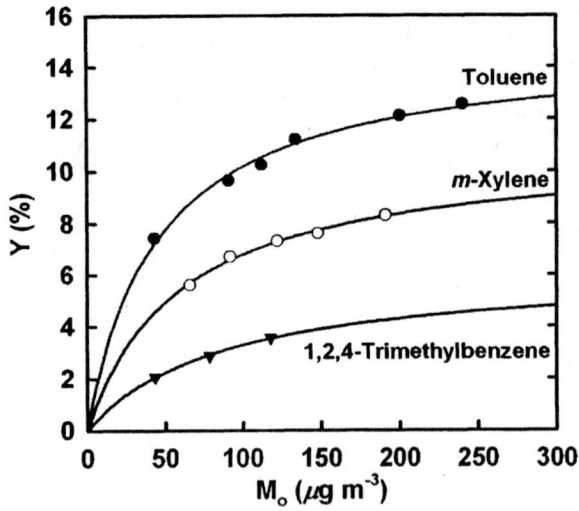

Figure 3. SOA yield (Y) variation with organic aerosol mass concentration (M_o) for seeds-free aromatic hydrocarbons/NO_x photooxidation experiments.

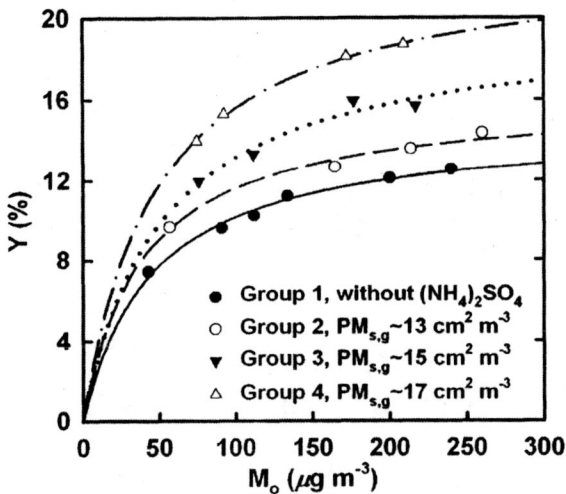

Figure 4. SOA yields variation with organic aerosol mass concentration for toluene/NO_x photooxidation experiments.

Table II. Summary of Initial Experiment Conditions and Results for Toluene/NO$_x$ Photooxidation Experiments. (303K, 58~62%RH)

Experiment No.	HC_0 (ppm)	NO_0 (ppb)	$NO_{x,0}$ (ppb)	PM_0 ($\mu m^3\ cm^{-3}$)	$\frac{HC_0}{NO_{x,0}}$	ΔHC ($\mu g\ m^{-3}$)	M_o ($\mu g\ m^{-3}$)	Y (%)
Tol-1	1.09	49	96	0	11.4	585	43	7.4
Tol-2	2.02	107	216	0	9.4	941	91	9.6
Tol-3	2.50	125	251	0	9.9	1090	112	10.2
Tol-4	2.85	143	289	0	9.9	1194	134	11.2
Tol-5	4.30	212	443	0	9.7	1655	200	12.1
Tol-6	5.10	250	511	0	10.0	1911	240	12.5
Tol-7	1.03	46	94	50	11.0	583	56	9.7
Tol-8	3.26	165	325	73	10.0	1297	164	12.7
Tol-9	4.14	209	442	72	9.4	1579	214	13.6
Tol-10	4.92	260	517	78	9.5	1816	260	14.3
Tol-11	1.24	56	118	67	10.5	634	76	12.0
Tol-12	1.93	105	208	74	9.3	840	112	13.3
Tol-13	2.52	122	244	65	10.3	1107	177	16.0
Tol-14	3.10	155	305	60	10.2	1382	217	15.7
Tol-15	1.04	49	102	75	10.2	534	75	13.9
Tol-16	1.41	70	142	85	9.9	604	92	15.3
Tol-17	2.01	104	210	70	9.6	951	172	18.1
Tol-18	2.51	123	245	66	10.2	1121	209	18.7

Table III. Information and Results of Categorization for Experiments with/without $(NH_4)_2SO_4$ Seed Aerosols

Group No.	Experiment No.	PM_0 ($\mu m^3\ cm^{-3}$)	$PM_{v.g}$ ($\mu m^3\ cm^{-3}$)	$PM_{s.g}$ ($cm^2\ m^{-3}$)	α	K_{om} ($\mu g^{-1}\ m^3$)	R^2
1	Tol-1~Tol-6	0	---	---	0.148	0.022	0.99
2	Tol-7	50	46	12.2	0.161	0.025	0.98
	Tol-8	73	57	12.9			
	Tol-9	72	62	13.2			
	Tol-10	78	57	13.2			
3	Tol-11	67	57	14.7	0.198	0.020	0.95
	Tol-12	74	68	15.3			
	Tol-13	65	59	15.7			
	Tol-14	60	56	14.4			
4	Tol-15	75	71	16.6	0.230	0.021	1.00
	Tol-16	85	82	18.5			
	Tol-17	70	66	17.4			
	Tol-18	66	62	17.1			

on the different $PM_{s,g}$ concentration. Group 1 includes results from particle-free experiments (Tol-1~Tol-6), while group 2 to 4 consist of those from experiments with $PM_{s,g}$ of about 13, 15 and 17 $cm^2 \ m^{-3}$, respectively. Using best-fit one-product model (i=1 in eq 1), four aerosol yield curves are obtained in Figure 4, and it indicates that aerosol yield increases with increasing particle surface concentration under the same organic aerosol mass concentration. $PM_{s,g}$ characterizes surface concentration of dry $(NH_4)_2SO_4$ aerosol when organic components start to partition onto the seed particles. Therefore, the dry $(NH_4)_2SO_4$ aerosol effect is believed to correlate with the variation of aerosol seed surface area.

The fitted values of α and K_{om} for each group are listed in Table III. The fitting results show that partitioning coefficient K_{om} has little change with the variation of $PM_{s,g}$ value, which implies that the overall condensability of condensable organic compounds (CCs) does not change. However, stoichiometric coefficient α increases with $PM_{s,g}$, which means that more aerosol-forming products (i.e. CCs) are generated with a higher surface concentration of $(NH_4)_2SO_4$ aerosol.

m-Xylene/NO_x and 1,2,4-Trimethylbenzene/NO_x Photooxidation Systems

The effect of $(NH_4)_2SO_4$ $PM_{s,g}$ values on SOA curves are also investigated for *m*-xylene and 1,2,4-trimethylbenzene (25). All experiments are conducted at 303K and about 60% RH. The hydrocarbon to NO_x ratios for *m*-xylene and 1,2,4-trimethylbenzene were fixed at about 6 and 8, respectively. The results are shown in Figure 5. Similar to toluene, the SOA yield data points can be categorized into several different groups based on the $PM_{s,g}$ values. In addition, the stoichiometric coefficient α increases with $PM_{s,g}$ increase, while the partitioning coefficient K_{om} is nearly unchanged. In the next section, we will provide one possible interpretation for the effect of $(NH_4)_2SO_4$ aerosol surface concentration on SOA formation.

Hypothesis of Dry $(NH_4)_2SO_4$ Seeds Effect

Unlike $(NH_4)_2SO_4$, it was shown that the presence of highly concentrated dry $CaSO_4$ and aqueous $Ca(NO_3)_2$ seed aerosols do not have significant effect on SOA formation (25, 26). After comparing the molecular composition of these three different inorganic compounds, we can find that both $CaSO_4$ and $Ca(NO_3)_2$ are neutral inorganic seeds, while $(NH_4)_2SO_4$ has some weak acidity, which might account for its effect on SOA formation. It was reported that organic compounds have little effect on the DRH of pure inorganic particles (27), which means the dry $(NH_4)_2SO_4$ particle will continue to be solid when it is covered by an organic layer. The organic layer contains both hydrophilic and hydrophobic

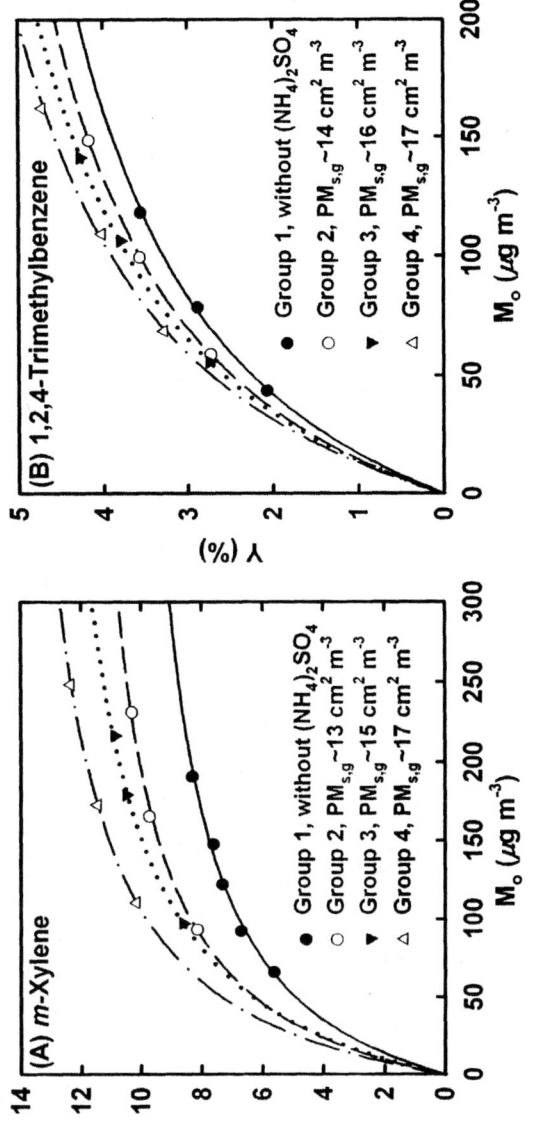

Figure 5. SOA yields variation with organic aerosol mass concentration for m-xylene/NO_x (A) and 1,2,4-trimethylbenzene/NO_x (B) photooxidation experiments.

species, and the hydrophilic species could absorb water molecules from the gas-phase, especially under high RH condition. Then, under the interaction of $(NH_4)_2SO_4$ and water-contained organic material, the aerosol surface may show some acidity due to the hydrolysis of NH_4^+.

It has been reported that strong acidic aqueous seeds can cause the heterogeneous reactions of carbonyl compounds, and enhance SOA yield by accelerating the formation of larger oligomers (*13, 14*). It is possible that the acidic $(NH_4)_2SO_4$ surface, similar to this strong acid effect, could also facilitate the acid-catalyzed heterogeneous reactions of carbonyl compounds, and produce oligomer from carbonyl monomer (*15*). The oligomerization of condensable organic compounds (CCs) will not change the amount of aerosol-forming products themselves, which means the mass-based stoichiometric coefficient α should not be changed. However, the optimized α increases with $PM_{s,g}$ (Table III), which indicates that some incondensable compounds (ICs, do not contribute SOA formation) could in some way produce aerosol-forming products in the presence of $(NH_4)_2SO_4$. We propose that some ICs can participate in the heterogeneous reactions occurring at the surface of $(NH_4)_2SO_4$, resulting in three types of oligomerization processes as shown in Table IV. The comprehensive results of these three types of reactions increase the amount of CCs produced, and therefore makes α higher.

The partition coefficient K_{om} was relatively unchanged through these four series of experiments, which can also be explained by the proposed ICs oligomerization. Both type 2 and type 3 oligomerization processes could produce a larger CC molecule, which lowers the volatility and increases the partition coefficient. However, type 1 process may generate CCs with smaller molecular weight and low condensability. The integrated results of these three different oligomerization processes can give a similar overall condensability (K_{om}) of CCs as when no acid-catalyzed oligomerization can take place.

Conclusions and Atmospheric Implications

Our results indicated that the presence of highly concentrated dry $(NH_4)_2SO_4$ aerosols has no significant effect on gas-phase reactions in aromatic

Table IV. Proposed Three Types of Oligomerization Processes

	Oligomerization Type	Effect on α	Effect on K_{om}
1	IC+IC+...	Increase α	Possibly decrease K_{om}
2	CC+CC+...	No effect on α	Increase K_{om}
3	IC+IC+...+CC+CC+...	Increase α	Increase K_{om}
	Integrated Result	Increase α	Possibly no change

hydrocarbon photooxidation systems, but enhances SOA generation and increases SOA yield. This effect is found to be positively related with the $(NH_4)_2SO_4$ surface concentration. Although the concentration of precursor VOC and NO_x in our experiments is several orders of magnitude higher than that in the ambient atmosphere, the effect of highly concentrated dry $(NH_4)_2SO_4$ aerosols is still expected to be prevalent in the atmosphere. The result of this study can be useful for SOA formation modeling, especially for air quality simulation modeling of developing cities experiencing serious PM pollution (e.g. Beijing).

Acknowledgements

This work was financially and technically supported by Toyota Motor Corporation and Toyota Central Research and Development Laboratories Inc. We would also like to thank Lanhua Hu for helpful discussions and Shan Wu for experiment preparation.

References

1. Jeffries, H. E. In *Composition, Chemistry, and Climate of the Atmosphere;* Singh, H. B., Ed.; Van Nostrand Reinhold: New York, 1995; pp 308-348.
2. Calvert, J. G.; Atkinson, R.; Becker, K. H.; Kamens, R. M.; Seinfeld, J. H.; Wallington, T. J.; Yarwood, G. *The Mechanisms of Atmospheric Oxidation of Aromatic Hydrocarbons;* Oxford University Press: New York, 2002.
3. Turpin, B. J.; Huntzicker, J. J. *Atmos. Environ.* **1995**, *29* (23), 3527-3544.
4. Duan, F. K.; He, K. B.; Ma, Y. L.; Jia, Y. T.; Yang, F. M.; Lei, Y.; Tanaka, S.; Okuta, T. *Chemosphere* **2005**, *60* (3), 355-364.
5. Eldering, A.; Cass, G. R. *J. Geophys. Res.-Atmos.* **1996**, *101* (D14), 19343-19369.
6. Pilinis, C.; Pandis, S. N.; Seinfeld, J. H. *J. Geophys. Res.-Atmos.* **1995**, *100* (D9), 18739-18754.
7. Cocker, D. R.; Clegg, S. L.; Flagan, R. C.; Seinfeld, J. H. *Atmos. Environ.* **2001**, *35* (35), 6049-6072.
8. Cocker, D. R.; Mader, B. T.; Kalberer, M.; Flagan, R. C.; Seinfeld, J. H. *Atmos. Environ.* **2001**, *35* (35), 6073-6085.
9. Kleindienst, T. E.; Smith, D. F.; Li, W.; Edney, E. O.; Driscoll, D. J.; Speer, R. E.; Weathers, W. S. *Atmos. Environ.* **1999**, *33* (22), 3669-3681.
10. Odum, J. R.; Hoffmann, T.; Bowman, F.; Collins, D.; Flagan, R. C.; Seinfeld, J. H. *Environ. Sci. Technol.* **1996**, *30* (8), 2580-2585.
11. Odum, J. R.; Jungkamp, T. P. W.; Griffin, R. J.; Forstner, H. J. L.; Flagan, R. C.; Seinfeld, J. H. *Environ. Sci. Technol.* **1997**, *31* (7), 1890-1897.

12. Takekawa, H.; Minoura, H.; Yamazaki, S. *Atmos. Environ.* **2003**, *37* (24), 3413-3424.
13. Gao, S.; Ng, N. L.; Keywood, M.; Varutbangkul, V.; Bahreini, R.; Nenes, A.; He, J. W.; Yoo, K. Y.; Beauchamp, J. L.; Hodyss, R. P.; Flagan, R. C.; Seinfeld, J. H. *Environ. Sci. Technol.* **2004**, *38* (24), 6582-6589.
14. Iinuma, Y.; Boge, O.; Gnauk, T.; Herrmann, H. *Atmos. Environ.* **2004**, *38* (5), 761-773.
15. Jang, M.; Carroll, B.; Chandramouli, B.; Kamens, R. M. *Environ. Sci. Technol.* **2003**, *37*, 3828-3837.
16. Jang, M.; Czoschke, N. M.; Lee, S.; Kamens, R. M. *Science* **2002**, *298*, 814-817.
17. He, K. B.; Yang, F. M.; Ma, Y. L.; Zhang, Q.; Yao, X. H.; Chan, C. K.; Cadle, S.; Chan, T.; Mulawa, P. *Atmos. Environ.* **2001**, *35* (29), 4959-4970.
18. Wang, Y.; Zhuang, G.; Sun, Y.; An, Z. *Atmos. Environ.* **2006**, *40* (34), 6579-6591.
19. Wu, S.; Lu, Z.; Hao, J.; Zhao, Z.; Li, J.; Takekawa, H.; Minoura, H.; Yasuda, A. *Adv. Atmos. Sci.* **2007**, *24* (2), 250-258.
20. Pandis, S. N.; Harley, R. A.; Cass, G. R.; Seinfeld, J. H. *Atmos. Environ.* **1992**, *26A* (13), 2269-2282.
21. Grosjean, D.; Seinfeld, J. H. *Atmos. Environ.* **1989**, *23*, 1733-1747.
22. Pandis, S. N.; Wexler, A. S.; Seinfeld, J. H. *Atmos. Environ.* **1993**, *24A*, 2403-2416.
23. Pankow, J. F. *Atmos. Environ.* **1994**, *28* (2), 185-188.
24. Pankow, J. F. *Atmos. Environ.* **1994**, *28* (2), 189-193.
25. Lu, Z.; Hao, J.; Li, J. *Atmos. Environ.* **2008**, *submitted*.
26. Lu, Z.; Hao, J.; Li, J.; Wu, S. *Acta Chim. Sinica.* **2008**, *66* (4), 419-423.
27. Cruz, C. N.; Pandis, S. N. *Environ. Sci. Technol.* **2000**, *34* (20), 4313-4319.

Chapter 9

Adsorption and UV Photooxidation of Gas-Phase Phenanthrene on Atmospheric Films

Jing Chen, Franz S. Ehrenhauser, Kalliat T. Valsaraj[*], and Mary J. Wornat

Cain Department of Chemical Engineering, Louisiana State University, Baton Rouge, LA 70803

The adsorption and UV photo-oxidation of gas-phase phenanthrene on atmospheric water films were studied using a flow-tube reactor. Bulk and interface air-water partition constants for phenanthrene were obtained. The interfacial partitioning of phenanthrene was increased in the presence of Suwannee River fulvic acid (SRFA) and sodium dodecyl sulfate (SDS). Three main photo-oxygenated products of phenanthrene were identified: 9, 10-phenanthrenequinone, 3,4-benzocoumarin, and 9-fluorenone. Photooxidation of phenanthrene proceeded faster as the film thickness decreased. Effects of SRFA and D_2O on the product formation rates were investigated. Based on the dependence of the product formation rates on the concentration of SRFA and D_2O, it was proposed that phenanthrene photodegraded via three different pathways: through radical cation intermediates, via reaction with singlet oxygen, and via reaction with hydroxyl radical.

© 2009 American Chemical Society

Introduction

Polycyclic aromatic hydrocarbons (PAHs) are ubiquitous contaminants in the environment (*1*). They range from the basic 2-ring naphthalene to multi-ring compounds of very large molecular weight. PAHs are considered persistent organic pollutants since they do not degrade easily in the environment unless external stimuli are given. The extensive π-orbital systems of PAHs allow them to absorb sunlight in the visible (400 – 700 nm) and the ultraviolet (290-400 nm) range of the solar spectrum (*2*). The photo-excitation of the PAH molecule often leads to the formation of oxygenated products which are more toxic than the parent compound (*3*). The parent compounds are hydrophobic and of low vapor pressure and they also display a tendency to adsorb at soil-water, sediment-water, and the air-water interfaces in the environment.

Atmospheric PAHs are subjected to a number of fate and transport processes through which their removal, distribution and transformations could occur. These processes can include physical removal by dry and wet deposition (rain, fog, snow), chemical and photochemical reactions in the gas phase and aerosol phase, and dispersion by convection. Whereas, the basic 2-ring compound, naphthalene is mostly present in the gas phase, phenanthrene and other higher molecular weight compounds reside primarily associated with the aerosol phase. The aerosols are composed of solid particles and liquid water as thin films. PAHs can be distributed between the solid and liquid phases of the aerosols. Thus the overall fate and transport of PAHs in aerosols are dependent on the processes occurring in both the solid and liquid aerosol fractions.

There have been numerous reports of the homogeneous reactions of gas-phase PAHs with atmospheric oxidants (*1*). Similarly, there have been some reports of the heterogeneous reactions of adsorbed PAHs with ozone, hydroxyl and nitrate radicals on solid particulate surfaces (soot, and aerosols) in the atmosphere (*4-6*). In general PAHs adsorbed to natural particles such as soot or fly-ash are more stable than in the pure form or adsorbed on silica gel, alumina or glass surfaces. Most of these reports are for dry particles in the atmosphere.

Air-water interface presents the largest environmental interface. This can be in the form of bulk phases in contact (air-sea), dispersed phases (air bubbles or water droplets) or thin films of water (aerosols). Apart from the equilibrium distribution of a chemical between bulk phases (water and air), very little information is available on the behavior at the air-water interface. When the surface area presented by water is much larger than the bulk volume, heterogeneous chemistry becomes more important than homogenous reactions in either the bulk air or water phases. Thin water films are ubiquitous in the atmospheric environment (e.g., aerosols, fog, ice) and for these the surface processes become more significant. It has been recently demonstrated from our laboratory via molecular dynamics simulations that the PAH molecules have deep free energy minima at the air-water interface (7). Experimentally it has been shown that the photoreactivity of PAH molecules at the air-water interface of a thin water film is enhanced as a result (8). Heterogeneous reactions of PAHs at the air-water interface with oxidants such as gas-phase ozone, singlet oxygen and hydroxyl radicals have been experimentally demonstrated (8-10). Reactions in thin films such as those on aerosols have been evaluated for only the most volatile naphthalene (8). For a few other PAHs (pyrene and anthracene) reactions in water-ice films and bulk water have been reported (4, 11, 12).

In this paper, we selected a semi-volatile compound (phenanthrene), a 3-ring PAH often detected in the ambient atmosphere, and studied its adsorption, photooxidation reaction and fate process at the air-water interface of water films. Phenanthrene is a persistent organic pollutant (POP) and is among the species monitored in the polar contaminant study (13). The behavior of phenanthrene is expected to be markedly different from naphthalene, which we reported earlier (8), since they differ in vapor pressure, aqueous solubility and hydrophobicity. Surfactant material such as long-chain alkanoic acids are often observed in aerosols and fog and they have been known to influence the behavior of gas-phase organic species in water films (14, 15). Hence, we also studied the effect of a natural surfactant film on the adsorption and photoreactivity of gas-phase phenanthrene in the water film.

Experimental

Materials

Chromosorb P (60-80 mesh size, acid washed) and porous polymer adsorbent (Orbo 43) were obtained from Supelco (Bellefonte, PA). Phenanthrene .(>96%), furfuryl alcohol (99%), 9, 10-phenanthrenequinone (>99%), 9- fluorenone (98%), and deuterium oxide (99.9% atom % D) were obtained from Aldrich. Suwannee River fulvic acid (SRFA) was obtained from the International Humic Substances Society (Cat. No. 1S101F). Sodium dodecyl sulfate (SDS) (≥99.5%) was obtained from Gibco BRL (Grand Island, NY). Water and acetonitrile used in HPLC were obtained from EMD Chemicals Inc. All above reagents were used as received. 3, 4-benzocoumarin was synthesized via Baeyer-Villiger oxidation of 9-fluorenone according to Mehta et al. (35) and purified by chromatography on an open silica column. The identity and purity (≥99%) were confirmed via HPLC-UV/MS and GC-MS by matching the mass spectral pattern (36).

Experimental Setup

Experiments were performed using a custom-constructed flow tube photoreactor described in detail in our previous work (*8, 16*). Phenanthrene vapor was generated by passing air through a PAH saturator and was then introduced to the flow tube photoreactor in which gas phase phenanthrene adsorbed onto a thin aqueous film coated on a 3.5 × 92 cm glass trough. The PAH saturator was made of six serially connected tubular columns (SS 316, 1/2" O. D., 0.37 m long). Each saturator column was packed with 15 g of Chromosorb P coated 10 percent by weight with phenanthrene. To obtain reproducible carrier gas flow rates in the PAH saturator, a mass flow controller (0-200 mL·min^{-1}, Aalborg Inc., Orangeburg, NY) was used to set the air flow rate to the saturator columns at 75 mL·min^{-1}. The gas phase concentration of phenanthrene obtained in this manner ranged from 0.4 to 0.7 µg·L^{-1} at room temperature (23 °C), while the reported value of the saturated gas phase concentration of phenanthrene ranges from 2.0 to 6.5 µg·L^{-1} (*17*).

Equilibrium Partitioning at the Air-water Interface

Uptake of phenanthrene onto thin aqueous films was investigated to study the interfacial behavior of phenanthrene at the air-water interface. The phenanthrene vapor/air mixture was introduced to the flow tube photoreactor through a moveable injector (SS 316, 1/8" O. D.) and adsorption of phenanthrene occurred on the aqueous film coated on the 3.5 × 92 cm glass trough. The temperature of the film was maintained at 296 K by a cooling bath. The aqueous sample was collected and the concentration of phenanthrene was quantified using a high performance liquid chromatograph (HPLC) after partition equilibrium was achieved between the gas and liquid phases and the aqueous concentration of phenanthrene ceased increasing. The time it took to reach equilibrium varied with the thickness of the aqueous film. To maintain consistency of our experiments, the adsorption duration was set as 10 hours, which is the time required to reach equilibrium on the thickest film (1714 µm) employed in our experiments. Adsorption of phenanthrene on pure water films, SRFA aqueous solution films and SDS aqueous solution films was investigated and the thickness of each kind of aqueous solution films ranged from 22 to 515 µm. As has been mentioned above, the gas phase concentration of phenanthrene generated by the PAH saturator varied from day to day between 0.4 and 0.7 µg·L^{-1}. In order to correct for the aqueous concentration variation brought about by the gas phase concentration change of phenanthrene, a pure water film of fixed thickness (1714 µm) was coated on a 3.5 × 5 cm glass trough and placed next to the target film, serving as the control for the adsorption experiments. The equilibrium concentration of phenanthrene in the target film was normalized to the equilibrium concentration in the control film for data analysis.

Measurement of Henry's Law Constant

The Henry's constant for phenanthrene was obtained by measuring the concentrations of phenanthrene in the bulk water and vapor phases in contact after equilibrium. A mini-bubbler (10 mL in capacity) filled with 5 mL deionized water was connected to the PAH saturator and the phenanthrene vapor/air mixture was bubbled through the bubbler at 75 mL·min^{-1}. After equilibrium was achieved between the water and vapor phases, a polymer trap was connected downstream to the bubbler to trap the vapor phase phenanthrene over a period of several hours. Water samples were withdrawn from the top of the bubbler both before and after collection of the vapor and analyzed using HPLC. The adsorbed phenanthrene was extracted into acetonitrile and also analyzed in HPLC. The vapor phase concentration of phenanthrene was estimated based on the gas flow rate, the vapor collection time, and the volume of acetonitrile used for extraction. Henry's constant was obtained by the direct ratio of the average aqueous concentration over the vapor collection period and the measured vapor phase concentration.

Photooxidation of Phenanthrene on Thin Aqueous Films

Photooxidation of phenanthrene on the thin aqueous film in the photoreactor was started after adsorption of phenanthrene onto the film was complete. Two UV lamps delivering UV light with wavelengths ranging between 280 and 315 nm were employed to provide illumination that simulated the UV-B component of sunlight. The UV light intensity on the surface of the aqueous film was 1.85 W·m^{-2}, around five times that of the UV-B solar irradiance at the Earth's surface for a midsummer day at 40 °N latitude (*38*). Photooxidation of phenanthrene was allowed to occur for a given duration of time before samples were taken for analysis in HPLC. Phenanthrene vapor was continuously introduced into the reactor throughout the experiments to compensate for the reacted aqueous phase phenanthrene. As a result, the aqueous concentration of phenanthrene remained constant. Effect of film thickness on the photooxidation rate of phenanthrene was investigated. The film thickness ranged from 22 to 515 µm. Effect of SRFA on the photooxidation rate of phenanthrene in a 515 µm aqueous film was also investigated and the concentration of SRFA ranged from 2 to 250 mg·L^{-1}.

Measurement of Singlet Oxygen

Singlet oxygen has been suggested to be the dominant reaction intermediate in the photooxidation of PAHs induced by UV light (*18, 19*). The steady state concentration of singlet oxygen in an illuminated water film with adsorbed phenanthrene was quantified by measuring the loss of low concentrations of furfuryl alcohol in illuminated 100% H_2O and 50/50 H_2O/D_2O films with adsorbed phenanthrene. The concentration change of furfuryl alcohol during illumination was determined using HPLC. The initial concentration of furfuryl alcohol used for the measurement was 1.2×10^{-6} M and the system studied was a 515 μm water film containing 3.5×10^{-6} M adsorbed phenanthrene.

Photooxidation of Phenanthrene on Aqueous Films Containing D_2O

The decay rate of singlet oxygen in dilute solution is controlled by solvent quenching and there is a considerable water-related H/D isotope effect on the lifetime of singlet oxygen. To investigate the role of singlet oxygen played in the photooxidation of phenanthrene, photooxidation rates of phenanthrene on 515 μm 100% H_2O, 50/50 H_2O/D_2O, and 100% D_2O films were measured.

Sample Analysis

Quantification of phenanthrene and photooxidation products in the aqueous samples was done using the same HPLC as was described in detail in our previous work (*16*). Identification of compounds was achieved by matching retention times of standard solutions within +/- 0.1 min and by matching the UV spectrum of the standards and the sample. The injection volume was 25 μL and the column thermostat was set to 40 °C. The mobile phase started with a 20/80 acetonitrile/water mixture and ramped to 80/20 acetonitrile/water within 12 min, then held at this concentration for 3 min, and finally returned to 20/80 acetonitrile/water in 3 min at a constant flow rate of 0.5 ml/min. The detection wavelength was set to 250 nm with 100 nm bandwidth and 4 nm slit.

Results and Discussion

Equilibrium Uptake of Gas-phase Phenanthrene on Water Films

Let us consider a film of water of thickness δ in the flow reactor that is exposed to a gaseous stream of phenanthrene at a constant concentration. The overall equilibrium uptake of phenanthrene in the water film is due to two processes, viz., adsorption at the air-water interface and dissolution within the bulk liquid. Thus, the overall equilibrium concentration of phenanthrene in the water film, C_w^T is given by

$$C_w^T = C_{w0} + \frac{C_{w0}}{K_{WA}} \frac{K_{\sigma A}}{\delta} \qquad (1)$$

Where K_{WA} is the dimensionless bulk water-air equilibrium partition constant (Henry's constant) for phenanthrene, $K_{\sigma A}$ (μm) is the partition constant at the air-water interface, and C_{w0} is the concentration in the bulk water phase for which the surface area is negligible in comparison to the bulk volume. The above equation clearly demonstrates that as the film thickness, δ is small, and the surface partition constant, $K_{\sigma A}$ is large, the contribution from the surface adsorption becomes larger.

Figure 1 shows the variation in the total aqueous concentration in the film as a function of the inverse of the water film thickness. A linear relationship was observed indicating the validity of the assumption that with decreasing film thickness (increasing $1/\delta$) the surface adsorption becomes the predominant uptake mechanism. The y-intercept and the slope of the plot give the values of the bulk phase uptake (C_{w0}) and $K_{\sigma A} \cdot C_{w0}/K_{WA}$ respectively. The bulk water-air equilibrium partition constant (K_{WA}) was determined separately by passing the phenanthrene vapor/air mixture through a bubbler and measuring the equilibrium concentrations of phenanthrene in the liquid and gas phases. The value determined was $K_{WA} = 1019$, which agreed with the value reported in other literature (Table I). From the y-intercept and the slope of Figure 1, one can, therefore calculate the value of the interface partition (adsorption) constant, $K_{\sigma A}$. The value of $K_{\sigma A}$ obtained was 3.3×10^4 μm and compared well with the estimate from correlation as shown in Table I. Using this value one estimates that at equilibrium 60% of phenanthrene will be present on the surface of an aqueous film 22 μm thick, whereas for a 515 μm film only 6% of the total mass of phenanthrene will be on the surface.

Figure 1 also shows that for an aqueous film that contains 207 mg.L^{-1} of SDS in the aqueous phase, which is equivalent to a monolayer of SDS, the partition constant, $K_{\sigma A}$ increases to 1.2×10^5 μm. The corresponding bulk water-air partition constant with aqueous phase SDS was 1019 showing no variation in the bulk phase equilibrium. Figure 1 also shows that for two different levels of SRFA in the aqueous phase, 51.5 and 280 mg.L^{-1}, which correspond to 15% and 50% surface coverage of SRFA (*16*), the values of $K_{\sigma A}$ obtained were 3.1×10^4 and 8.1×10^4 μm respectively. The corresponding K_{WA} values for the two aqueous concentrations of SRFA were 1019 and 1175 respectively indicating little variation in the bulk phase equilibrium. Thus, it is clear that the presence of surface active materials in the aqueous phase at substantial concentrations effectively increase the overall equilibrium partitioning to the air-water interface and uptake by the water film. The degree of increase in the equilibrium partitioning was determined by the surface coverage of the surfactants and the hydrophobic interactions between the surfactants and phenanthrene.

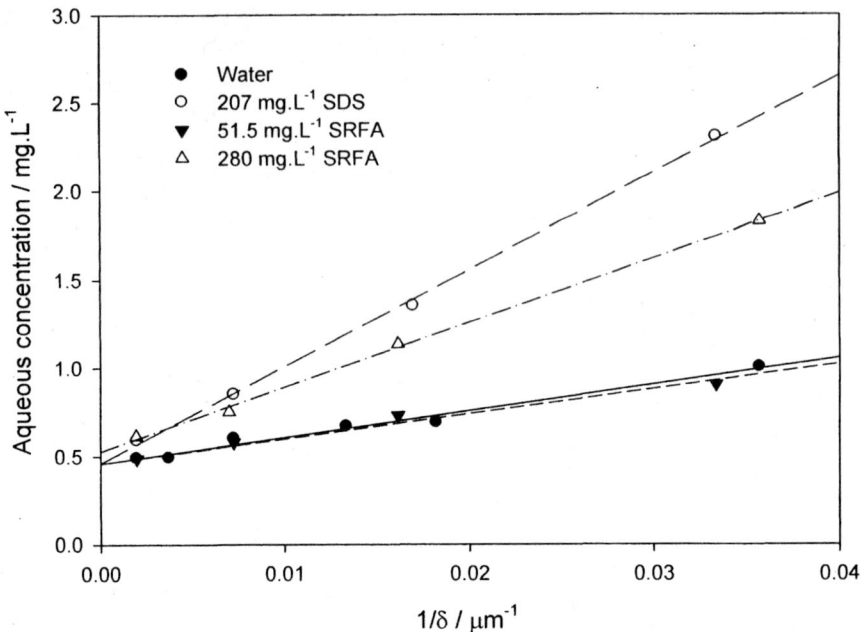

Figure 1. Uptake of phenanthrene from the gas phase on aqueous films with varying thicknesses.

Table I. Bulk and Interface Air-water Partition Constants of Phenanthrene (T=296 K)

Partition constant	Water		SDS 207 mg.L^{-1}	SRFA 51.5 mg.L^{-1}	SRFA 280 mg.L^{-1}
	This work	Reference			
K_{WA} /[-]	1019	955 (17)	1019	1019	1175
$K_{\sigma A}$ /[μm]	3.3 x 10^4	3.5 x 10$^{4\,a}$	1.2 x 10^5	3.1 x 10^4	8.1 x 10^4

[a] Data obtained from correlation: log ($K_{\sigma A}$/m) = +0.940 log (K_{OA}/[-]) − 8.607; r^2 = 0.987 (20). K_{OA} is the octanol-air partition constant for the compound. log K_{OA}=7.602 for phenanthrene at 298 K (21).

Photochemical Reactions of Phenanthrene on Thin Aqueous Films

Photochemical reactions of phenanthrene under simulated sunlight conditions were studied after adsorption of phenanthrene onto the film was complete. Figure 2 shows the typical HPLC trace of an aqueous film containing adsorbed phenanthrene that was exposed to UV radiation for 12 hours. The chromatogram shows several peaks apart from phenanthrene, among which three compounds with a relatively high abundance were identified and confirmed with pure standards. The products identified were: 9, 10-phenanthrenequinone, 9-fluorenone and 3, 4-benzocoumarin. Note that in the case of the thin film experiments reported in this work we see the three main products in all of our samples.

Figure 3 shows the accumulation of the three main products in a 515μm water film as the reaction proceeds. Quantification of the products was done on HPLC.

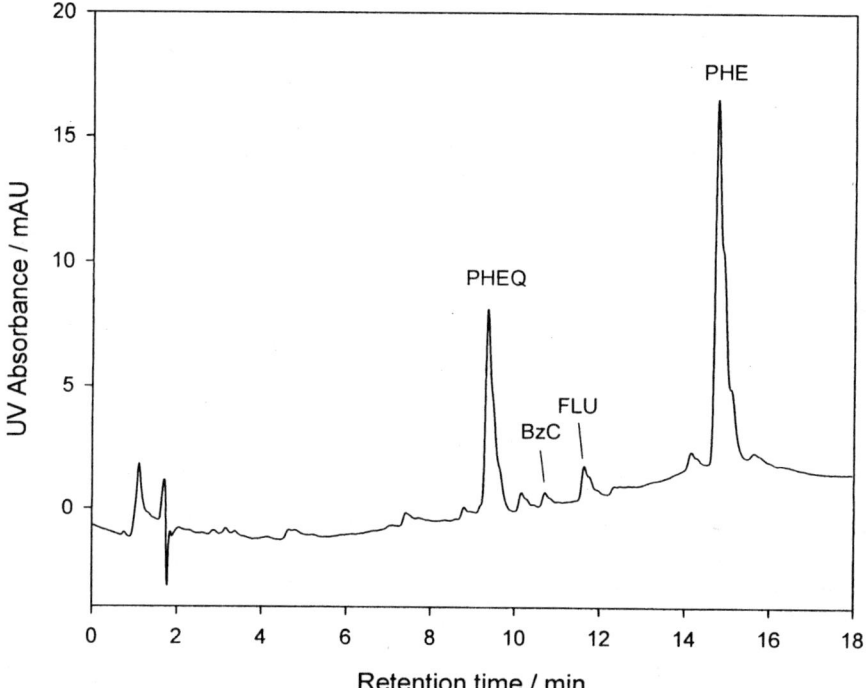

Figure 2. HPLC trace of a 515 μm aqueous film sample after 12 hours exposure to UV light. The compounds identified are: 9, 10-phenanthrenequinone (PHEQ), 3, 4-benzocoumarin (BzC), 9-fluorenone (FLU), and phenanthrene (PHE).

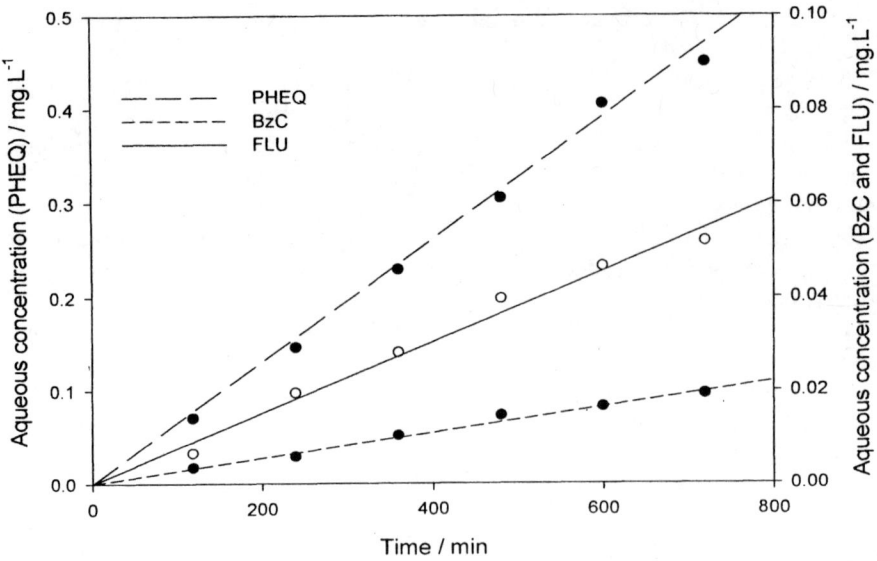

Figure 3. Accumulation of the three main photooxidation products of phenanthrene in a 515µm water film.

An overall reaction scheme, $PHE \xrightarrow{k_1} P1 \xrightarrow{k_2} P2$, was proposed to interpret the kinetic data. As a result, the concentration change of the products during the reaction can be described as

$$C_{P1}(t) = \frac{k_1}{k_2} \cdot C_{Phe0}\left(1 - e^{-k_2 t}\right) \quad (2)$$

where C_{P1} is the concentration of the product, C_{Phe0} is the concentration of phenanthrene, k_1 is the product formation rate constant, and k_2 is the product reaction rate constant. Note that the concentration of phenanthrene, C_{Phe0}, was kept constant throughout the reaction because of the continuous supply of phenanthrene from the gas phase. Moreover, since the water-air equilibrium partition constants are high for the oxygenated product compounds, e.g., $K_{WA} = 3.6 \times 10^4$ for 9-fluorenone at 298K (37), we can neglect their gas phase concentrations within the reactor. Separate control experiments with standards of products dissolved in water also showed that the evaporative loss of products from the water film in the reactor was negligible. A detailed deduction of Eqn. 2 was given in our previous work (8). In the case where P1 is a stable product, further degradation of P1 can be neglected and thereby Eqn. 2 can be simplified to

$$C_{P1}(t) = k_1 C_{Phe0} t \quad (3)$$

As shown in Figure 3, the concentrations of the products increase linearly with time and the kinetic data fits well to Eqn. 3. The formation rate constants for the products were obtained by fitting the kinetic data to Eqn. 3.

Effect of Film Thickness on Reaction Rates

Figure 4 shows that the formation rate constants for the three main products from phenanthrene greatly increased as the film thickness decreased. Table II lists the product formation rate constants measured in water films with both the largest (515 μm) and the smallest thickness (22 μm) employed in our experiments. Increases of 47% (9, 10-phenanthrenequinone), 1495% (3, 4-benzocoumarin), and 1264% (9-fluorenone) were observed as the film thickness decreased from 515 μm to 22 μm. Heterogeneous reactions at the gas-aqueous interface have been shown to proceed faster than homogeneous reactions in the bulk water phase (8, 22, 23). The contribution of surface reaction increases as the surface area per unit volume (1/δ) increases, therefore the measured overall formation rate constants for the products increased with decreasing film thickness. Similar effect of film thickness on the photooxidation rate of naphthalene was shown in our previous work (8). However, the highest increase in the product formation rate constant for naphthalene was 154% as the film thickness decreased to 22 μm, compared to 1495% for phenanthrene. The reported values of $K_{\sigma A}$ and K_{WA} for naphthalene were 21 μm and 86 respectively (8). Using these values it can be calculated that at equilibrium only 1% of naphthalene is present on the surface of an aqueous film 22 μm thick, whereas 60% of phenanthrene is present on the same aqueous film. Given a fixed film thickness, the proportion of surface reaction for phenanthrene is higher than for naphthalene, therefore, the measured overall rates for product formation from phenanthrene show a higher increase.

Effect of SRFA on Reaction Rates

It is shown in literature reports that fulvic acid is the most important component of dissolved organic matters in natural waters. It is also shown via molecular characterization that Suwannee River fulvic acid (SRFA) is a good surrogate model to represent polycarboxylic acids in fog waters (14). In this work, we chose SRFA to study the effect of dissolved surfactants on the photooxidation of phenanthrene in thin water films. Figure 5 shows the effect of SRFA on the observed formation rate constants for 9, 10-phenanthrenequinone, 3, 4-benzocoumarin, and 9-fluorenone in a 515 μm aqueous film. Interestingly, the effects of SRFA on the observed formation rate constants for the three main products were totally different. The formation rate constant of 9, 10-

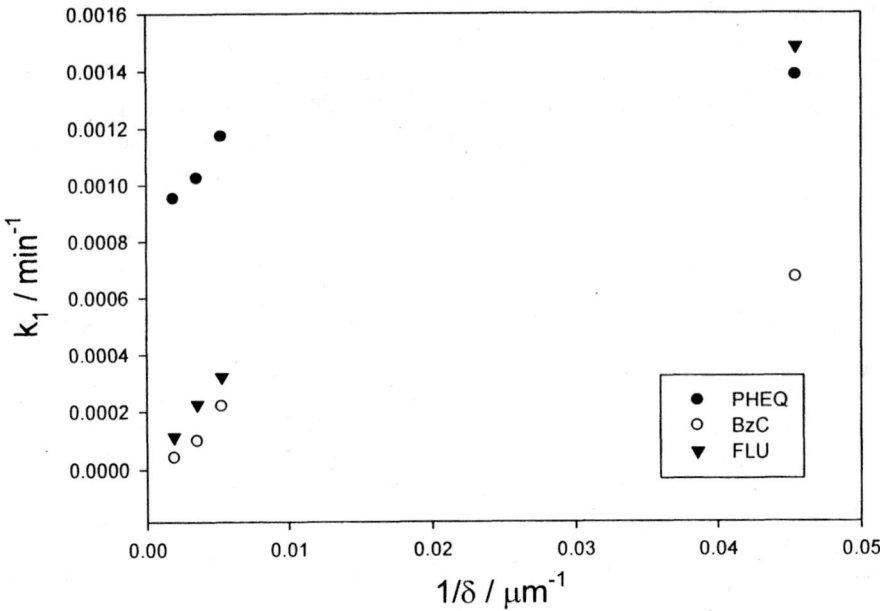

Figure 4. Effect of film thickness on the observed formation rate constants of 9,10-phenanthrenequinone, 3,4-benzocoumarin, and 9-fluorenone.

Table II. Observed Product Formation Rate Constants in Water Films (T=296 K)

Compound	k_1 / min^{-1} (515 μm film)	k_1 / min^{-1} (22 μm film)	Percentage increase
PHEQ	9.5×10^{-4}	1.4×10^{-3}	47%
BzC	4.2×10^{-5}	6.7×10^{-4}	1495%
FLU	1.1×10^{-4}	1.5×10^{-3}	1264%

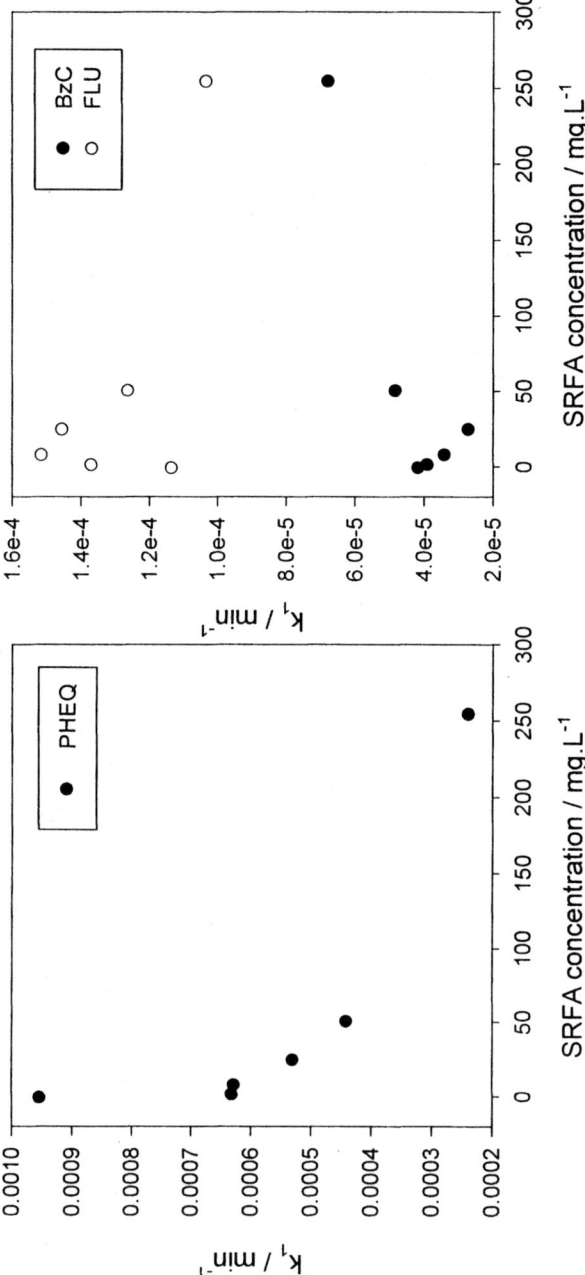

Figure 5. Effect of SRFA on the observed formation rate constants of 9, 10-phenanthrenequinone, 3, 4-benzocoumarin, and 9-fluorenone in a 515 μm aqueous film.

phenanthrenequinone decreased monotonically as the concentration of SRFA increased, whereas that of 3, 4-benzocoumarin decreased in the beginning and then increased. Contrary to 3, 4-benzocoumarin, the formation rate constant of 9-fluorenone increased first and then decreased. Literature reports on the effect of SRFA on PAH photodegradation in bulk water have appeared quite conflicting. Fasnacht and Blough reported that photoreactivities of PAHs in bulk water solutions were not affected by SRFA (*24*). However, other reports showed that whereas the photodegradation of benzo[a]pyrene and benzo[a]anthracene were slowed by humic-like substances in water, that of naphthalene increased (*25*). These seemingly conflicting results are probably attributed to the different reaction mechanisms that different PAHs undergo. In the case of phenanthrene photooxidation in our work, the different effects of SRFA on the formation of the three main products suggest that phenanthrene was photooxidized to the products via different pathways. Effect of SRFA on phenanthrene photooxidation is discussed further in the following section.

Photooxidation Pathways of Phenanthrene

Three main pathways via which PAHs photodegrade have been proposed in the literature: through radical cation intermediates, via reaction with singlet oxygen ($^1O_2$), and via reaction with hydroxyl radical ($\cdot OH$) (*24, 26*). Figure 6 shows the three possible pathways of PAH photooxidation in oxygen-containing water. It has been proposed in the literature that singlet oxygen is the dominant reaction intermediate in the direct photooxidation of PAHs induced by UV light (*18, 19*). In our work, we used furfuryl alcohol (FFA), an efficient $^1O_2$-selective trapping agent, to quantify the steady state concentration of singlet oxygen in the illuminated 515 μm water film with adsorbed phenanthrene. The steady state concentration of singlet oxygen was determined by (*27*)

$$[^1O_2]_{ss} = \frac{k_{exp,D_2O} - k_{exp,H_2O}}{k_{FFA,^1O_2}[\frac{k_{H_2O}}{k_{H_2O}\chi_{H_2O} + k_{D_2O}\chi_{D_2O}} - 1]} \qquad (4)$$

Detailed description of Eqn. 4 and values of the constants can be found in the references (*27, 28*). The apparent first-order kinetic rate constants for the loss of FFA in 100% H_2O and 50/50 H_2O/D_2O films with adsorbed phenanthene (k_{exp,H_2O} and k_{exp,D_2O}) were determined to be 2.0×10^{-5} s^{-1} and 2.5×10^{-5} s^{-1} respectively. Knowing all the values of the parameters on the right-hand side of Eqn. 4, the steady state concentration of singlet oxygen in the illuminated 515 μm water film with adsorbed phenanthrene can be calculated and the value determined was 4.9×10^{-14} M.

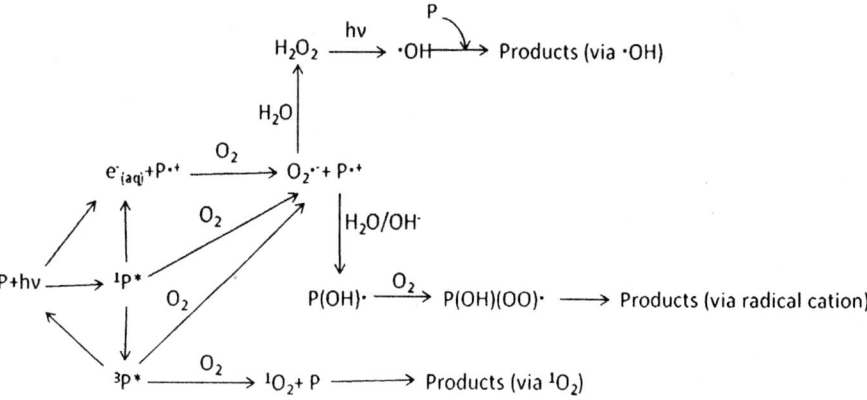

Figure 6. Scheme of photooxidation pathways of PAHs in the O_2/H_2O system.

The decay rate of singlet oxygen in dilute solution is controlled by solvent quenching. Water-related deuterium isotope shows a considerable effect on the lifetime of singlet oxygen and there is evidence in the literature that the lifetime of singlet oxygen is longer in D_2O than in H_2O (*29*). Therefore, it is anticipated that the photooxidation of phenanthrene would proceed faster in D_2O. However, as shown in Figure 7, 3, 4-benzocoumarin was the only one of the three main products that had an increasing formation rate constant as the amount of D_2O in the film increased, confirming its formation route via singlet oxygen. Unlike 3, 4-benzocoumarin, the formation rate constant of 9, 10-phenanthrenequinone decreased with increasing D_2O and that of 9-flurenone increased first and then decreased. The different effects of D_2O on the formation rate constants of the three products further confirmed that phenanthrene was photooxidized to the products via different pathways. D_2O is a more ordered liquid than H_2O, thus the lifetimes of reactive species are longer in D_2O. However, D_2O exhibits stronger hydrogen bonds than H_2O, which makes it more difficult to generate deuteroxyl radical while illuminated by UV light. Literature reports also show that the yields of radicals and molecular products in the radiolysis of D_2O are lower than in H_2O (*30, 31*). Therefore, the concentration of hydroxyl or deuteroxyl radical in the illuminated water film with adsorbed phenanthrene decreases with increasing amount of D_2O thus the photooxidation process of phenanthrene via the hydroxyl radical pathway becomes slower. The effect of D_2O on the formation rate constant of 9, 10-phenanthrenequinone coincided with this phenomenon, suggesting that 9, 10-phenanthrenequinone was formed via the hydroxyl radical pathway. As shown in Figure 6, photooxidation of phenanthrene via the radical cation pathway involves both water and reactive species transformed from phenanthrene. Reactivity of water decreases with increasing D_2O, whereas the lifetime of reactive species transformed from phenanthrene increases. The two opposite effects of D_2O would result in a complex effect of D_2O on phenanthrene photooxidation via the radical cation pathway. It is hypothesized that 9-fluorenone was formed from phenanthrene via the radical cation pathway.

Now let us look back at the effects of SRFA on phenanthrene photooxidation. SRFA can quench or scavenge the PAH excited states, free radicals, or other excited species that may be intermediates in the photochemical reactions of PAHs. Therefore, the formation rate constant of 9, 10-phenanthrenequinone, which was produced via the hydroxyl radical pathway, decreased with SRFA. On the other hand, UV-light absorption of fulvic acid molecules can also promote them to their singlet excited state and generate singlet oxygen via the same route as PAHs (32-34). It was shown in our previous work that photooxidation rate of naphthalene via the singlet oxygen pathway decreased first and then increased with SRFA (16), which was identical to the case of 3, 4-benzocoumarin. The effect of SRFA on the formation of 9-fluorenone is more complex than the other two and needs further investigation.

Comparison of Reaction Rates via Radical Cations, $^1O_2$, and ·OH

As shown in Table II, the formation rate constant of 9, 10-phenanthrenequinone is one order of magnitude higher than that of 9-fluorenone and 3, 4-benzocoumarin in the 515 μm water film, indicating that photooxidation via hydroxyl radical is the most favorable reaction pathway for phenanthrene in bulk water under the simulated sunlight conditions employed in our experiments. However, the formation rate constants of 9-fluorenone and 3, 4-benzocoumarin increased much faster than that of 9, 10-phenanthrenequinone as the film thicknesses decreased. The formation rate constant of 9-fluorenone was even higher than that of 9, 10-phenanthrenequinone in the 22 μm water film. Therefore, it can be concluded that photooxidation via the radical cation and $^1O_2$ pathways are more favorable than via the hydroxyl radical pathway for the surface reaction of phenanthrene.

Conclusions

Surface processes are significant for the transport and transformations of PAHs in the atmospheric condensed phases (e.g., aerosols, fog and cloud droplets). It is fairly well documented that the behavior of pollutants in fogs is extremely size dependent. The study on the adsorption and UV photo-oxidation of gas-phase phenanthrene on atmospheric water films was intended to provide insight into the behavior of PAHs in the atmospheric environment where thin

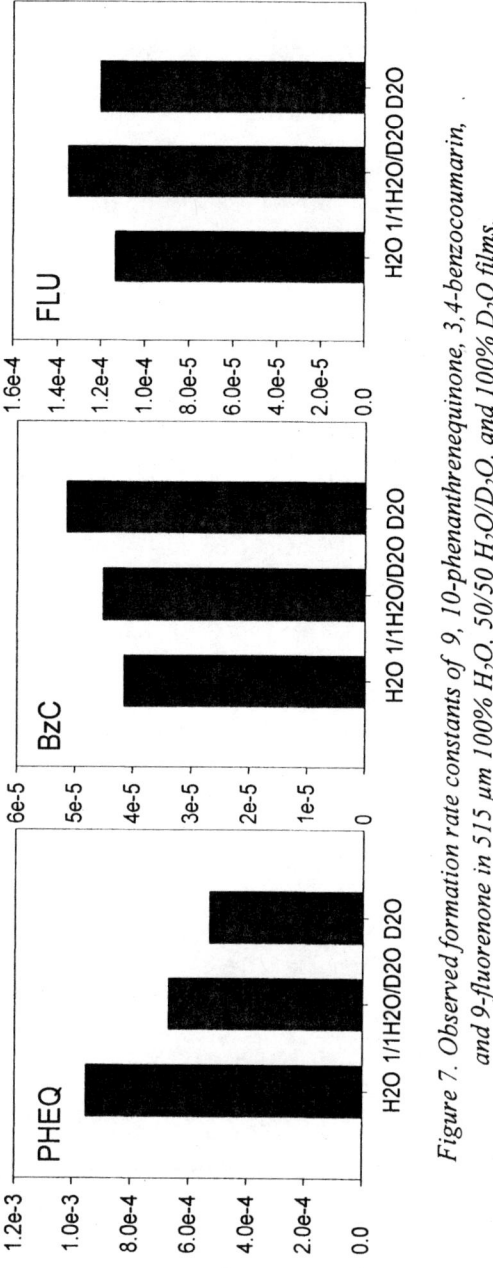

Figure 7. Observed formation rate constants of 9, 10-phenanthrenequinone, 3,4-benzocoumarin, and 9-fluorenone in 515 μm 100% H_2O, 50/50 H_2O/D_2O, and 100% D_2O films.

water films are ubiquitous. Completion of this study enabled us to draw the following conclusions.

The total aqueous concentration of adsorbed phenanthrene increased linearly with the surface area per unit volume ($1/\delta$) of the water film. Bulk and interface air-water partition constants for phenanthrene can be determined by measuring the dependence of the total aqueous concentration of adsorbed phenanthrene on $1/\delta$. The interfacial air-water partition constant of phenanthrene increased greatly in the presence of SRFA and SDS, indicating that the presence of surface active materials in the aqueous phase at substantial concentrations effectively increased the equilibrium partitioning to the air-water interface and uptake by the water film.

9, 10--phenanthrenequinone, 3, 4-bezocoumarin and 9-fluorenone were identified as the three main photooxidation products of phenanthrene. The surface reaction of phenanthrene proceeded faster than the reaction in bulk water. Phenanthrene photodegraded through three different pathways under simulated sunlight conditions: through radical cation intermediates, via reaction with singlet oxygen ($^1O_2$), and via reaction with hydroxyl radical ($\cdot OH$). Photooxidation via $\cdot OH$ was the most favorable reaction pathway for phenanthrene in bulk water, whereas photooxidation via the radical cation and $^1O_2$ pathways were more favorable for the surface reaction.

Acknowledgement

This work was supported by a grant from the National Science Foundation (ATM 0355291).

References

1. Finlayson-Pitts, B. J., Pitts, J. N. *Chemistry of the Upper and Lower Atmosphere*; Academic Press: San Diego, CA, 2000.
2. Nikolaou, K., Masclet, P., Mouvier, G. *Sci. Total Environ.* **1984**, *32*, 103-132.
3. McConkey, B. J., Duxbury, C.L., Dixon, D.G., Greenberg, B.M. *Environ. Toxicol. Chem.* **1997**, *16*, 892-899.
4. Kahan, T. F., Donaldson, D. J. *J. Phys. Chem. A* **2007**, *111*, 1277-1285.
5. Kwamena, N.-O. A., Thornton, J.A., Abbatt, J.P.D. *J. Phys. Chem. A* **2004**, *108*, 11626-11634.
6. Perraudin, E., Budzinski, H., Villenave, E. *Atmos. Environ.* **2007**, *41*, 6005-6017.
7. Vacha, R., Jungwirth, P., Chen, J., Valsaraj, K. T. *Phys. Chem. Chem. Phys.* **2006**, *8*, 4461-4467.

8. Chen, J., Ehrenhauser, F. S., Valsaraj, K. T., Wornat, M. J. *J. Phys. Chem. A* **2006**, *110*, 9161-9168.
9. Raja, S., Valsaraj, K. T. *J. Air & Waste Manage. Assoc.* **2005**, *55*, 1345-1355.
10. Mmereki, B. T., Donaldson, D. J., Gilman, J. B., Eliason, T. L., Vaida, V. *Atmos. Environ.* **2004**, *38*, 6091-6103.
11. Bernstein, M. P., Sandford, S. A., Allamandola, L. J., Gillette, J. S., Clemett, S. J., and Zare, R. N. *Science* **1999**, *283*, 1135-1138.
12. Domine, F., Albert, M., Huthwelker, T., Jacobi, H.-W., Kokhanovsky, A. A., Lehning, M., Picard, G., and Simpson, W. R. *Atmos. Chem. Phys. Discuss.* **2007**, *7*, 5941-6036.
13. Hung, H. H., Blanchard, P., Halsall, C. J., Bidleman T. F., Stern, G. A., Fellin, P., Muir, D. C. G., Barrie, L. A., Jantunen, L. M., Helm, P. A., Ma, J., Konoplev, A. *Sci. Total Environ.* **2005**, *342*, 119-144.
14. Cappiello, A., De Simoni, E., Fiorucci, C., Mangani, F., Palma, P., Trufelli, H., Decesari, S., Facchini, M. C., Fuzzi, S. *Environ. Sci. Technol.* **2003**, *37*, 1229-1240.
15. Latif, M. T., Brimblecombe, P. *Environ. Sci. Technol.* **2004**, *38*, 6501-6506.
16. Chen, J., Valsaraj, K.T. *J. Phys. Chem. A* **2007**, *111*, 4289-4296.
17. Montgomery, J. H. *Groundwater Chemicals: Desk Reference*, 2nd ed.; CRC Press/Lewis Publishers: Boca Raton, FL, 1996.
18. Vione, D., Maurino, V., Minero, C., Pelizzetti, E., Harrison., M. A. J., Olariu., R. I., Arsene., C. *Chem. Soc. Rev.* **2006**, *35*, 441-453.
19. Suzdorf, A. R., Morozov, S. V., Kuzubova, L. I., Anshits, N. N., Anshits, A. G. *Chem. Sust. Dev.* **1994**, *2-3*, 449-473.
20. Valsaraj, K. T. *Environ. Toxicol. Chem.* **2004**, *23*, 2318-2323.
21. Li, X., Chen, J., Zhang, L., Qiao, X., Huang, L. *J. Phys. Chem. Ref. Data* **2006**, *35*, 1365-1384.
22. Nissenson, P., Knox, C. J. H., Finlayson-Pitts, B. J., Phillips, L. F., Dabdub, D. *Phys. Chem. Chem. Phys.* **2006**, *8*, 4700-4710.
23. Remorov, R. G., George, C. *Phys. Chem. Chem. Phys.* **2006**, *8*, 4897-4901.
24. Fasnacht, M. P., Blough, N. V. *Environ. Sci. Technol.* **2002**, *36*, 4364-4369.
25. Bertilsson, S., Widenfalk, A. *Hydrobiologia* **2002**, *469*, 23-32.
26. Miller, J. S., Olejnik, D. *Wat. Res.* **2001**, *35*, 233-243.
27. Anastasio, C., McGregor, K. G. *Atmos. Environ.* **2001**, *35*, 1079-1089.
28. Allen, J. M., Gossett, C. J., Allen, S. K. *J. Photochem. Photobiol. B: Biol.* **1996**, *32*, 33-37.
29. Kajiwara, T., Kearns, D. R. *J. Am. Chem. Soc.* **1973**, *95*, 5886-5890.
30. Hayon, E. *J. Phys. Chem.* **1965**, *69*, 2628-2632.
31. Gauduel, Y., Pommeret, S., Antonetti, A. *J. Phys. Chem.* **1993**, *97*, 134-142.
32. Zepp, R. G., Schlotzhauer, P. F., Sink, R. M. *Environ. Sci. Technol.* **1985**, *19*, 74-81.
33. Haag, W. R., Hoigne, J. *Environ. Sci. Technol.* **1986**, *20*, 341-348.

34. Latch, D. E., McNeill, K. *Science* **2006**, *311*, 1743-1747.
35. Mehta, G., Pandey, P. N. *Synthesis* **1975**, *6*, 404-5.
36. Patel, J. R., Politzer, I. R., Griffin, G. W., Laseter, J. L. *Biomed. Mass Spectrom.* **1978**, *5(12)*, 664-70.
37. Meyer, D., Dressler, H., Ruck, W. *LCGC North America* **2007**, *25(2)*, 180-190.
38. Zepp, R. G., Cline, D. M. *Environ. Sci. Technol.* **1977**, *11*, 359-366.

Aerosol Modeling

Chapter 10

Understanding Climatic Effects of Aerosols: Modeling Radiative Effects of Aerosols

Tarek Ayash, Sunling Gong, and Charles Q. Jia

Department of Chemical Engineering and Applied Chemistry, University of Toronto, Toronto, Ontario M5S 3E5, Canada

Climate on the Earth is a highly dynamic and complex system in which aerosols have been increasingly recognized as a key component. While the production, transport and fate of aerosols are fundamentally determined by the elements of climate, such as wind and and precipitation, aerosols may affect the Earth's climate through complex processes of absorbing and reflecting the incoming solar and the outgoing terrestrial radiation, and indirectly affecting solar and terrestrial radiation by changing the properties of clouds, in addition to participating in heterogeneous reactions that affect key atmospheric constituents. Due to inherent complexities, coupled with modeling limitations, quantifying the aerosols' climate effects is still highly uncertain and, thus, presents a challenging aspect of climate research. In this chapter, the elements of arosol-climate interactions and the uncertainties underlying aerosol-climate modeling are reviewed. Climatic implications of radiative forcing are discussed.

Introduction

The direct effect of aerosols involves the scattering and absorption of solar radiation by the aerosol particles. It is critically determined by some of their intrinsic properties, and is also influenced by certain atmospheric conditions. Aerosols' interference with solar radiation is dictated by their radiative properties, which in turn is fundamentally determined by the particle's size, shape and chemical composition. The interaction of aerosols with solar radiation is greatest when the aerosol dimensions are similar to the radiation wavelength. Hence, the longer lived accumulation mode (0.1–1 μm) influences the shortwave solar radiation spectrum to a greater extent than the longwave terrestrial infrared spectrum. At these longer infrared wavelengths, the low residence time of cfoarse-mode aerosols is considered to result in a smaller radiative interaction (1). Increasing evidence suggests that the shape of many atmospheric aerosols is far from being spherical, and can sometimes be highly asymmetrical. Unlike cloud droplets, which scatter radiation symmetrically in every direction, aerosols shape may significantly influence their scattering behavior, resulting in scattered radiation peaking in certain directions. Optical properties of aerosols also depend on their chemical composition, which determines their index of refraction. As such, aerosols can be either absorbers or non-absorbers of solar/terrestrial radiation.

Further complications are caused by the reflectivity of underlying surfaces, termed "albedo". Over surfaces with low albedo, such as the oceans, an aerosol layer tend to backscatter solar radiation, thereby tending to increase the planetary albedo. In contrast, the scattering caused by an aerosol layer situated above a highly reflecting surface, such as snow, can result in a light-trapping effect, which lowers the net albedo. In addition, the direct radiative effect of aerosols depends on the type and altitude of clouds (2-4). When a cloud layer is present above aerosols, most of the incident radiation will be reflected back to space and only a small fraction will interact with aerosols. On the other hand when an elevated aerosol layer is present with a cloud below, the aerosols interact not only with radiation incident from the Sun, but also with that reflected from the cloud layer below. This results in an enhanced aerosol radiative impact. Overall, the interplay of many factors determines the magnitude of the aerosol's direct effect, and may even alter the sign of the forcing (positive or negative).

The indirect effects of atmospheric aerosols refer to the possible modification of cloud properties, through the alteration of the physicochemical properties of cloud condensation nuclei (CCN) and ice nuclei (IN). By acting as CCN, therefore, aerosols may impact the radiative properties of clouds basically in two ways. First, increasing the number of cloud condensation nuclei (CCN) leads to more but smaller cloud droplets in a cloud (whose liquid water content remains constant), which enhances the cloud's reflection of solar radiation. This is known as the first indirect effect or the Twomey effect. In addition, increased number of smaller cloud droplets reduce the precipitation efficiency and therefore enhance the cloud lifetime and hence the cloud reflectivity, which is referred to as the cloud lifetime or second indirect effect. Moreover, a semi-direct effect occurs due to the absorption of solar radiation by aerosols, leading to a heating of the air, which can result in evaporation of cloud droplets (5). This warming can partially offset the cooling due to the indirect aerosol effect, and both the cloud lifetime effect and the semi-direct effect involve feedbacks because the cloud lifetime and cloud liquid water content change. An increase in the number of ice crystals in cirrus clouds would also exert a Twomey effect in the same way that the Twomey effect acts for water clouds. In addition, a change in the ice water content of cirrus clouds could exert a small radiative effect in the infrared range (6).

Aerosol and Climate Modeling

From a modeling perspective, direct radiative forcing by aerosols can be accurately calculated, at least in principle, once the optical constants, size distribution, and atmospheric concentration of the aerosol are known. Accurate determination of the indirect aerosol effect is more problematic since this involves complex physical interactions that are not fully understood. The difficult part is predicting and modeling the time evolution of the climate system's response in reaction to the imposed forcing. This is because changes in temperature caused directly by the imposed radiative forcing produce feedback reactions within the climate system that tend to magnify and geographically shift the direct effect of the forcing (7). Despite substantial recent advancements in measurement and modeling tools, aerosol and climate modeling in general, and aerosol-climate interactions in particular are still posing great challenges for the modeling community. The elements of, up-to-date findings and the uncertainties underlying such modeling are reviewed in this chapter.

Elements of and Approaches in Modeling

Elements of aerosol-climate modeling may be grouped into four categories: climate elements, aerosol cycling, aerosol optics and radiation transfer. Possible connections and dependencies among these categories are illustrated in Figure 1.

Climate Elements

Although it is desirable to know the general state of climate, certain climate elements are more essential to aerosol modeling than others. These include meteorological elements of wind, humidity, cloud cover and precipitation;

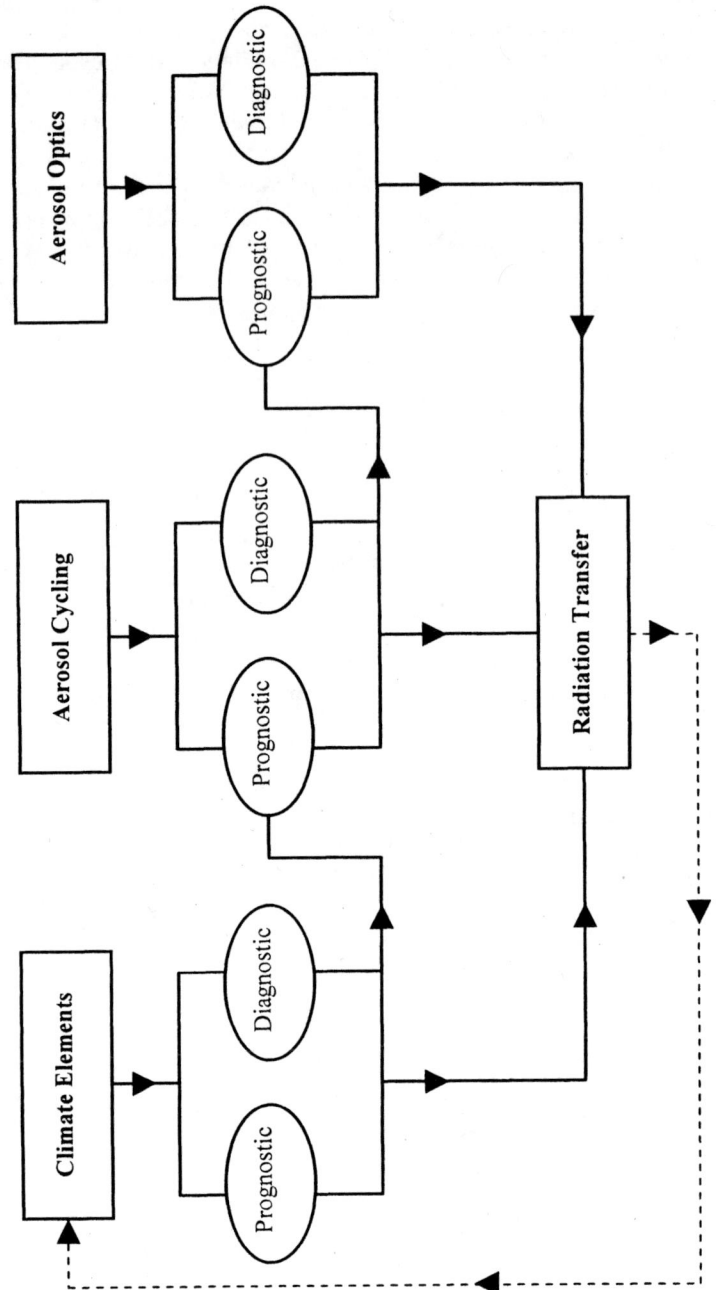

Figure 1. Connections and dependencies among elements of modeling aerosol-climate interactions.

surface elements of roughness and land cover; and radiative elements of shortwave and longwave fluxes, and surface albedo. These fields may be diagnosed by using climatological-mean values that are obtained either from measurement records or from model-generated datasets. Prognostically, these elements may be computed by sophisticated global climate models.

A global climate model is the most comprehensive tool available for studying the importance of any forcing on the climate system, including aerosol. Such a model consists of a number of components: an atmospheric general circulation model (AGCM), some form of an ocean model, a sea ice model, and a land surface model. AGCMs are numerical representations of the equations of motion and physical processes that define the workings of the atmosphere. The spatial resolution of AGCMs is usually determined by computational resources. Physical processes included in AGCMs are radiation, convection, and processes in the boundary and surface layers. Cloud processes related to cloud amount and optical properties are typically included as a part of the radiation calculations. All of these processes operate on scales smaller than those resolvable in the AGCM and hence require parameterization. These cannot always be well constrained, because observations on the micro-scale are lacking.

Aerosol Cycling

Aerosol cycling refers to elements that define the emission, transport, transform, fate and physicochemical characteristics of aerosols. These elements may be diagnosed mainly through field sampling campaigns or by remote sensing using satellites and lidar instruments. Measuring the global, three-dimensional distributions of aerosol parameters desired for climate models is theoretically possible but prohibitively expensive. Satellites can determine the global distribution of a limited number of aerosol properties integrated over a vertical column, with some additional information on the vertical distribution; however, they cannot be used to determine the chemical composition of the particles. Surface and aircraft-based platforms can provide much of the very detailed information needed, but with limited coverage in space and time.

Prognostically, some or all of these elements may be computed through models that vary in complexity, the most comprehensive of which being an Atmospheric Chemical Transport Model (ACTM). Predictions of aerosol distributions by an ACTM require information on magnitudes and geographic distributions of emissions of precursor gases and sources of primary particles, chemical reaction rates in the atmosphere, transport of these gases and aerosols by large-scale advection and subgrid-scale convection, and removal mechanisms. "Online" ACTMs are linked to an AGCM through large-scale advection of gases and aerosols, which employs atmospheric winds, and through convection parameterization, which supplies convective fluxes for subgrid vertical transport. Links are also associated with the land and surface modules in climate models through the representation of dry deposition processes. Wet removal processes in the ACTM require precipitation information from the AGCM. Finally, aqueous-phase reactions require information of cloud water. "Off-line" ACTMs employ either AGCM atmospheric state information or meteorological or climatological state information.

Aerosol size is an important determinant of optical properties, cloud nucleating properties, and wet and dry removal rates. The processes determining particle size distribution include: direct injection of primary particles, nucleation of new particles, condensation, coagulation, hygroscopic growth, heterogeneous production in clouds, and mixing of different air masses. Several approaches are employed to describe the particle size distributions in numerical models. The simplest cases are the so-called bulk schemes, where the sizes of the aerosol particles are constant, and only the aerosol mass is predicted. In modal schemes, the particle size distribution is represented by mathematical functions. In the third type, called bin (or spectral) schemes, the aerosol size distribution is represented by several size intervals. The accuracy along with the computational costs increases with the number of bins for which the aerosol mass is predicted.

In dealing with ACTMs, a primary uncertainty is related to the inadequate specification of source rates. Furthermore, to test model predictions against measurements, it is necessary that natural components as well as anthropogenic components be represented. Direct aerosol forcing requires accurate specification of particle mass distribution, whereas indirect aerosol forcing requires accurate knowledge of particle number distribution. Accurate representation of aerosol forcing is more likely if chemical species are treated separately rather than trying to represent aerosols according to regional types such as maritime, continental, or stratospheric. The use of several size modes for a given aerosol species may lead to more accurate representation of the prognostic evolution of size distributions in climate models (*8*).

Aerosol Optics

Aerosol's interference with atmospheric radiation, and hence their climatic effect, is dictated by their optical properties. Prognostically, they can be determined based on some "fundamental" properties, from which other parameters can be derived. Alternatively, prescribed or measured optical properties can be input into radiative models. Recently, advancements in satellite and ground-based remote sensing made possible the retrieval of some optical properties with wide spatial coverage. Such data have been actually used in some modeling studies of aerosol direct radiative forcing (*9*); however, full reliance on these data for modeling purposes remains hindered by limitations in the retrieval algorithms, spatiotemporal extent and frequency of measurements, and high uncertainties in or the absence of certain parameters, such as the factor accounting for multiple-scattering of photons and the a-priori assumption of aerosol size distribution.

Knowing the chemical composition of the aerosol, its optical parameters are fundamentally determined by its real and imaginary indices of refraction over the full range of visible and thermal wavelengths. According to the International Union of Pure and Applied Chemistry, the imaginary part of the complex index of refraction represents the absorption index and the real part represents the refractive index. Hence, the former gives a measure of the particle's radiative absorptivity, while the latter indicates the particle's ability to scatter radiation. Optical constants can be readily measured in the laboratory for pure chemical compounds that are known to be constituents of atmospheric aerosols. However, there are both fundamental and practical problems in obtaining reliable optical constants and radiative parameters for a given atmospheric aerosol. For example, it is not correct to average the optical constants obtained for pure compounds to simulate the complicated mix of chemical compounds and impurities that characterize typical aerosols, and laboratory results remain sample-specific and are not directly applicable to aerosols with a different mix of impurities. Also, there is a related problem in being able to model the effect on aerosol refractive indices and radiative parameters caused by the change in composition and particle size for hygroscopic particles as they respond to changes in relative humidity. Therefore, the task of defining a reliable refractive index database is made more difficult by the need to distinguish external and internal mixtures of different chemical species.

Given the wavelength-dependent index of refraction and the particle size distribution, the radiative parameters that fully describe the radiative properties of aerosols can then be calculated. The basic quantities needed to describe the direct interaction of aerosol particles with solar radiation are the aerosol optical depth, single-scattering albedo, and angular scattering function. The aerosol optical depth is the vertical integral of the aerosol extinction coefficient – a measure of the particles' ability to absorb and scatter radiation. The single-scattering albedo is a measure of the relative magnitudes of aerosol scattering and absorption strengths, and is defined as the ratio of the scattering to extinction (scattering + absorption) coefficients. The angular scattering function describes the angular distribution of the intensity of radiation scattered by particles. Parameterizations in radiative transfer models use an angle-integral property of the angular scattering function, either the asymmetry factor or the upscatter fraction.

Mie theory permits calculations of aerosol radiative properties; however, it is based on the assumption that particles are spherical (which may be acceptable for hygroscopic particles, but may be poor for the hydrophobic ones). Also, the state of mixture of different chemical species must be known. For internally-mixed particles, which contain two or more chemical species, mixture rules for calculating the mixture's refractive index exist but must be evaluated against observational and laboratory measurement data, particularly for the case where insoluble, light-absorbing material is mixed with soluble, nonabsorbing species.

Unlike cloud droplets, which scatter radiation symmetrically in every direction, aerosols shape may significantly influence their scattering behavior, resulting in scattered radiation being rather peaked in certain directions. Detailed theoretical comparisons of the radiative properties of nonspherical polydispersions and Mie-scattering results (*10,11*) show that nonspherical particle effects can be large and important. For climate modeling applications, where only radiative flux and albedo information are required, Mie-scattering calculations made for equivalent volume spheres appear to give adequate accuracy, provided that the optical depth, size distribution, and refractive index of the aerosol are known a priori (*7*).

Radiation Transfer

Knowing the aerosol radiative properties and the necessary climatic radiative elements, radiative transfer models are used to calculate the aerosol direct effect. The indirect aerosol effects are also computed by incorporating into the radiative transfer model the aerosol-modified cloud radiative properties, which are more or less equivalent to those of aerosols (given that both are commonly treated as spheres in radiative models). To that end, radiative models range in complexity from simple equations, to one layer box-models, up to multi-layer radiative transfer schemes. For example, highly simplistic equations have been developed that give a first-order estimate of the aerosol radiative forcing, by calculating the change in planetary albedo resulting from the addition of the aerosol layer uniformly over the globe (*12*). However, to properly quantify the aerosol radiative effects, more sophisticated treatments of the atmospheric radiative transfer problem are needed. In addition, realistic estimates for the radiative forcing by aerosols require that the radiative calculations be performed in the context of a global atmosphere that includes overlapping absorption by ozone, water vapor, and carbon dioxide, with surface, cloud, and Rayleigh scattering under clear and cloudy sky conditions. In the context of three-dimensional climate modeling, radiation transfer in a vertically heterogeneous atmosphere is commonly treated following a multi-layer approach, where fluxes are computed by calculating the absorption and scattering of solar and terrestrial radiation assuming homogeneous, parallel layers. Because of the strong solar zenith angle dependence of multiple scattering for small optical depths, accurate results require the use of fairly rigorous and numerically expensive computational methods.

Finally, to account for the climate effects of aerosols, their radiative effects must be incorporated into a climate model. To fully account for aerosol-climate interactions and feedbacks, however, their modeling components must be fully coupled; whereby aerosol's direct and indirect radiative effects, as well as their effect on cloud lifetimes, are fed back into a driver climate model.

The state of the art and uncertainties

Aerosol cycling and optical properties

In 2003, the aerosol model inter-comparison initiative-AeroCom was created to provide a platform for detailed evaluations of aerosol simulation in global models, where current global aerosol simulations based on harmonized diagnostics are analyzed (*13,14*). Based on the AeroCom-models-average results (Table 1), emissions are found to be dominated in mass by sea-salt (SS), followed by dust, sulfate, particulate organic matter (POM), and black carbon (BC); while burdens (total aerosol mass in the atmosphere) from the greatest to the least are: dust, SS, sulfate, POM and BC. Residence times are found to be longest for BC, followed by POM, dust, sulfate, and sea-salt; BC, POM and especially sulfate reach greater heights than other components. Sulfate, BC and POM are found to be predominantly removed by wet deposition, while about two-thirds of dust and sea-salt are removed by dry deposition.

The diversities established in the AeroCom comparison indicate that aerosol processes in the atmosphere are still not completely understood. Several processes and parameters, which are particularly relevant for aerosol radiative forcing calculations, with high diversities are: masses of aerosol in the radiatively active fine mode, dry aerosol composition, aerosol water content, and vertical aerosol dispersal. Consequently, the improved representation of these processes and parameters in large-scale aerosol models deserves a high priority in order to reduce the uncertainty of the climatic impact attributed to aerosol.

AeroCom analyzed model median aerosol optical thickness (AOT, a combined measure of the vertical extent of aerosol layer and the radiative extinction (absorption plus scattering) efficiency of its particles) fields for total, component and absorption contributions, as well as co-single-scattering albedo (Co-SSA, or one minus SSA, where SSA is a measure of the scattering efficiency of aerosol particles and defined by the ratio of scattering to extinction efficiencies). Results are presented in Table 2. Sulfate clearly dominates the AOT contribution over industrialized regions. Dust's contribution dominates over the desert regions, and sea-salt contributes significantly over the remote southern-ocean regions. The modeled absorption potential is strongest in the tropical biomass regions, with a seasonal peak which occurs prior to the seasonal peak for AOT. Also the absorption potential is larger for Europe than for Asia or North America. Relative low is the absorption potential for the Eastern U.S., while lowest values are modeled for ocean regions away from sources.

Table 1. AeroCom statistics of selected aerosol life-cycle parameters

Parameter (unit)	Dust		SS		Sulphate	
	Mean	SD (%)	Mean	SD (%)	Mean	SD (%)
Emission (Tg/yr)	1,840	49	16,600	199	179	22
Burden (Tg)	19.2	20.5	7.52	54	1.99	25
Residence time (days)	4.14	43	0.48	58	4.12	18
Total removal rate (1/day)	0.31	62	5.07	188	0.25	18
Wet removal rate (1/day)	0.08	42	0.79	77	0.22	22
Dry removal rate (1/day)	0.23	84	4.28	219	0.03	55
Parameter (unit)	BC		POM		All Aerosols	
	Mean	SD (%)	Mean	SD (%)	Mean	SD (%)
Emission (Tg/yr)	11.9	23	96.6	26	18,800	176
Burden (Tg)	0.24	42	1.7	27	30.6	29
Residence time (days)	7.12	33	6.54	27	1.42	65
Total removal rate (1/day)	0.15	21	0.16	24	2.27	223
Wet removal rate (1/day)	0.12	31	0.14	32	0.3	64
Dry removal rate (1/day)	0.03	55	0.03	49	1.98	250

SOURCE: Reproduced with permission from reference 13. Copyright 2006.

Table 2. AeroCom annual global statistics for mid-visible AOT

AOT_{550nm}	Median	Max-Min Ratio
Sulfate	0.034	3.4
BC	0.004	5.2
POM	0.019	5.0
Dust	0.032	4.5
SS	0.030	3.3
Total	0.127	2.3
Absorption	0.005	3.2

SOURCE: Reproduced with permission from reference 14. Copyright 2006.

The agreement among models improved for the global annual AOT during the last couple of years. Most simulated global averages now agree well with consolidated high-quality data from remote sensing. However, model diversity for each of the five component AOT contributions individually is significantly larger than for the combined total AOT (*15*). Model diversity usually increases towards remote regions, suggesting that aerosol processing during long-range transport is a key issue for reductions of model diversity. Model diversity is usually larger over land than over oceans for total dry mass and total AOT. The largest differences occur in central Asia and extend eastwards to western regions of North America. Diversity for aerosol absorption is significantly larger than diversity for aerosol optical thickness.

Radiative forcing

The definition of forcing is restricted to changes in the radiation balance of the Earth-troposphere system imposed by external factors, with no changes in stratospheric dynamics, without any surface and tropospheric feedbacks in operation, and with no dynamically-induced changes in the amount and distribution of atmospheric water (*16*). Applied to atmospheric aerosol, therefore, the term "radiative forcing" could only refer to anthropogenic changes in the state of aerosols, between pre-industrial and current times, either by adding to their atmospheric loading (e.g. through industrial activities) or by inducing changes in the loading of natural aerosols (e.g. desertification increasing soil dust).

According to the 4th IPCC report (*15*), the total direct aerosol forcing as derived from models and observations is estimated to be -0.5 [± 0.4] W m^{-2}, with a medium-low level of scientific understanding. The direct forcing of the individual aerosol species is less certain than the total direct aerosol forcing. The estimates are: sulphate, -0.4 [± 0.2] W m^{-2}; fossil fuel organic carbon, -0.05 [± 0.05] W m^{-2}; fossil fuel black carbon, $+0.2$ [± 0.15] W m^{-2}; biomass burning, $+0.03$ [± 0.12] W m^{-2}; and mineral dust, -0.1 [± 0.2] W m^{-2}. The uncertainties remain large, and rely to a large extent on the estimates from global modeling studies that are difficult to verify. Three major areas of uncertainty exist: uncertainties in the atmospheric burden and the anthropogenic contribution to it, uncertainties in the optical parameters, and uncertainties in implementation of the optical parameters and burden to give a radiative forcing. The problem of the degree of external/internal mixing in the atmosphere deserves highlighting, as global modeling studies tend to assume external mixtures which make modeling simpler.

Of all the aerosol radiative forcings, the indirect aerosol effect is the most complicated and difficult to determine and, hence, the most uncertain. The forcing due to the cloud albedo effect in the context of liquid water clouds (Twomey effect) is estimated to be -0.7 Wm^{-2}, with a low level of scientific understanding (15). (16) reviewed forcing estimates of the aerosol indirect effects published since the 2001 IPCC report (Table 3).

The upper negative bound is slightly reduced from -2.0 to -1.85 Wm^{-2}. All the models suggest a negative Twomey effect with the smallest one being -0.22 Wm^{-2}. Beyond the Twomey effect, both the cloud lifetime effect and the semi-direct effect involve feedbacks because the cloud lifetime and cloud liquid water content change. Therefore, they were not included in the IPCC's radiative forcing assessment. The cloud lifetime effect is estimated to be roughly as large as the Twomey effect. Through the semi-direct effect, absorption of solar radiation by aerosols leads to a heating of the air, which can result in an evaporation of cloud droplets. This warming may partially offset the cooling due to the indirect aerosol effect. Conversely, the semi-direct effect can also result in a cooling depending on the location of the black carbon with respect to the cloud.

Table 3. Estimates of the aerosol indirect forcing (Wm^{-2}) published since the 2001 IPPC report

Effect	TOA	Surface
Indirect aerosol effect for clouds with fixed water amounts (Twomey effect)	-0.5 to -1.9	Similar to TOA
Indirect aerosol effect with varying water amounts (cloud lifetime effect)	-0.3 to -1.4	Similar to TOA
Semi-direct effect	$+0.1$ to -0.5	Larger than TOA

SOURCE: Reproduced with permission from reference 16. Copyright 2005.

Radiative effect

While "radiative forcing" mainly refers to the changes induced by anthropogenic emissions, a common modeling interest is to quantify the extent to which the presence of aerosols, natural and anthropogenic, affects the Earth-atmosphere's radiative budget in the current state of climate. This is known as the aerosol "radiative effect" and applies to all components of the atmospheric aerosol. Estimates of the direct radiative effect of the combined and individual aerosol species are listed in Table 4. Different models show an overall cooling effect of the combined tropospheric aerosols. While (17) and (18) use modeled aerosol properties, (9) and (19) follow a different approach by using values of aerosol properties that are derived from remote sensing. The high-end value obtained by (9) could be explained in part by the fact that it is for oceanic regions only, where the predominant sea-salt and sulfate aerosols are highly scattering. The wide range of estimates of the direct effect of the sea salt aerosol

Table 4. Estimates of the TOA, global, annual-mean direct radiative effect (Wm^{-2}) due to different aerosol species.

Species	Clear sky	Whole sky	Reference
Sea salt	-1.51 to -5.03 [1]		(20) [2]
	-0.59	-0.31	(17) [3]
	-1.1	-0.54	(18)
	-0.15		(24) [2]
	-2.2	-1.1	(25)
	-1.50	-0.65	(21) [2]
Dust	+0.26	+0.36	(17)
	-0.23	-0.14	(18)
	+0.07		(22)
	-0.11		(26)
	-0.18		(27)
	-0.4		(28)
	-1.12	-0.60	(29)
Sulfate	-0.72	-0.40	(17)
	-0.84	-0.44	(18)
BC	+0.21	+0.36	(17)
		+0.2 ±0.15	(15)
OC	-0.45	-0.24	(17)
	-0.12	-0.07	(18)
All	-1.29	-0.24	(17)
	-2.48	-1.23	(18)
	-5.4		(9) [1,2]
	-2.2		(19) [2]

[1] Computed over oceans only
[2] Solar radiative effect only
[3] Computed at the tropopause

mainly reflects the high uncertainty underlying its emission modeling. This is due to the fact that the generation mechanisms of sea salt are highly determined by wind speed. Hence, uncertainties in modeling wind speeds, added to limitations in the parameterizations of sea salt's generation mechanisms, are reflected in estimates of the aerosol radiative effect. This is well-demonstrated by (20), who report "low" (-1.51 Wm^{-2}) and "high" (-5.03 Wm^{-2}) estimates of sea-salt direct effect by using two different empirical approaches that encompass the considerable range in the effect of wind speed on sea-salt aerosol concentration. The recent work by (21) was based on the Canadian Aerosol Module (CAM), which relies on climatological input from the Third Generation Canadian Climate Center General Circulation Model (CCC GCM III) (21).

While considerable uncertainty underlies the magnitude of the sea-salt direct effect, even the sign of the dust's effect remains undetermined. In addition to the problem associated with the shortwave-longwave competing effects of dust, major uncertainty relates to the fact that the radiative effects of aerosols depend strongly on the single scattering albedo, which in turn depends on the real and imaginary components of the aerosol refractive index. For dust, these radiative properties are highly dependent on particle size and mineralogical composition, which are also interrelated for minerals (22). The use of a single spectrum of refractive indices for all atmospheric dust involves considerable approximation, given that the measured values of the imaginary refractive index spans nearly an order of magnitude over large sections of both the shortwave and longwave spectral regions (23).

Climatic implications

A growing number of model applications supported by observations suggest a role of aerosols in affecting the climate through atmospheric heating and cooling, which results in altering atmospheric circulation and precipitation patterns at various spatial and temporal scales. Based on GCM climate simulations, (30) suggested that interactions of greenhouse gas forcing plus direct and indirect aerosol effects could explain why surface observations show puzzling evidence of reduced solar warming and concurrent increasing temperature during the last four decades. They argued that, while reductions in surface solar radiation due to clouds and aerosols are only partly offset by enhanced down-welling longwave radiation from the warmer and moister atmosphere, the radiative imbalance at the surface leads to weaker latent and sensible heat fluxes and hence to reductions in evaporation and precipitation despite global warming.

Takemura et al. (31) simulate the changes in the meteorological parameters of clouds, precipitation, and temperature caused by the direct and indirect forcings of aerosols between pre-industrial (1850) and present (2000) days. Their model results indicate that a decrease in the cloud droplet effective radius of about 10% due to the first indirect effect by anthropogenic aerosols occurs globally, while changes in the cloud water and precipitation are strongly affected by a variation of the dynamical hydrological cycle with a temperature change by the aerosol direct and first indirect effects rather than the second indirect effect itself. However, their results also show that the cloud water can increase and the precipitation can simultaneously decrease in regions where a large amount of anthropogenic aerosols and cloud water exist, which can be seen as a strong signal of the second indirect effect. Their study suggests that aerosol particles approximately reduce 40% of the increase in the surface air temperature by anthropogenic greenhouse gases on the global mean.

Model simulations by (*32*) show that aerosol cooling extends up to the tropopause with a maximum in the boundary layer of the northern mid and high latitudes. In the tropics aerosol cooling is at maximum in the upper troposphere. The overall effect of the aerosol forcing is a cooling near the surface in the polluted regions of the Northern Hemisphere that stabilizes the lower atmosphere whereas the near surface changes in temperature are smaller in the tropics and the mid-latitudes of the Southern Hemisphere. A destabilization of the atmosphere above the boundary as a result of black carbon heating within the boundary layer was obtained in a climate model study by (*33*). Their GCM simulations also demonstrated that absorbing aerosols (mainly black carbon) heat the air, alter regional atmospheric stability and vertical motions, and affect the large-scale circulation and hydrologic cycle with significant climate effects. That helped them to better explain why there has been a tendency in recent decades toward increased summer floods in South China, increased droughts in north China, and moderate cooling in China and India while most of the world has been warming.

Over the Indian Ocean region during the dry winter monsoon season it has been estimated that anthropogenic aerosols especially the highly absorbing aerosols can decrease the average solar radiation absorbed by the surface in the range of 15 to 35 Wm^{-2} (*34*). This results in an increase in the atmospheric heating between the surface and 3 km altitude by up to 60 to 100%. Similar perturbations in the atmosphere have been observed over other regions namely East Asia, South America, sub-Saharan Africa, which are subjected to large loading of absorbing aerosols. Such a perturbation imposed over the Indian Ocean in the 15°S – 40°N and 50 – 120°E region can lead to a large regional cooling at the surface in the range of 0.5 to 1K accompanied by a warming of the lower troposphere by about 1K as has been deduced from a GCM study with fixed sea surface temperatures (*35*). This vertical heating gradient alters the latitudinal and inter-hemispheric gradients in solar heating and these gradients play a prominent role in driving the tropical circulation (*36*) and determining the amount of precipitation (*37*).

Using a global climate model/mixed-layer ocean model (*38,39*) showed that the dynamical and hydrological changes in the Sahel region (North Africa) in response to the indirect effect of anthropogenic sulfate aerosols are similar to the observed changes that have been associated with the Sahelian drought for the period 1900 – 1998. The modeled anthropogenic aerosol cooling dominates on the Northern Hemisphere, which causes a southward shift of the intertropical convergence zone. If, on the contrary, the Northern Hemisphere surface temperature is increased more than the Southern Hemisphere surface temperature due to the increase in fossil fuel combustion of black carbon, then the intertropical convergence zone shifts northward, strengthens the Indian summer monsoon and increases the rainfall in the Sahel (*40*).

A shift of the ITCZ, northward however, was also attributed to absorbing haze, which consists mainly of anthropogenic aerosols and spans most of south Asia and the north Indian Ocean in dry seasons between November and May (35). The modeled dynamical response to the aerosol forcing in recent years was found surprisingly large, and was caused principally by concurrent cooling of the land surface and warming of the atmosphere, and a consequent weakening of the north-south temperature gradient in the lower level. Beyond local effects, (41) went as far as showing that the South-Asian haze forcing and its fluctuation could explain some of the recent observed boreal-wintertime changes of the southwest Asian monsoon, El Nino-Southern Oscillation (ENSO) and the Arctic Oscillation. They base their claims on the facts that their simulations reveal a wintertime drought over southwest Asia, a poleward shift of the Northern Hemisphere (NH) zonal-mean zonal momentum during the winter season and a slightly equatorward shift of the NH extratropical zonal momentum, as well as a significant suppression of the convection in the western equatorial Pacific during the boreal wintertime that would weaken the trade winds over the Pacific and induce warm anomalies in the eastern basin.

In addition to anthropogenic aerosols, soil dust can also interfere with atmospheric circulation. Through radiative forcing at the surface, dust alters the hydrologic cycle, since the reduction in sunlight beneath the dust layer is balanced globally by diminished evaporation, with a secondary reduction in the surface sensible heat flux (42). Simulating the current climate by an AGCM, (43) find dust to reduce precipitation by only a percent or two. However, for glacial climates, when the inferred load is substantially larger (44), they estimate that the reduction in precipitation could exceed 5%, roughly half of the precipitation decrease resulting simply from the colder climate (45), suggesting that dust radiative forcing contributes non-negligibly to the reduction of the hydrologic cycle during glacial climates. The reduction in rainfall also has the potential to reduce vegetation, thus expanding potential regions of dust emission. Despite its global reduction, (43) find that evaporation is increased along with rainfall by dust over the Sahara. They suggest that this could be explained by the fact that, over the Sahara, a modest heating anomaly by dust radiative forcing can change the sign of the total diabatic heating, resulting in ascent and precipitation.

References

1. Baltensperger, U.; Nyeki, S. In *Physical and chemical properties of aerosols*; Colbeck, I., Ed.; Blackie Academic and Professional: London, 1998, pp 281-330.
2. Heintzenberg, J.; Charlson, R.; Clarke, A.; Liousse, C. R., V; Shine, K.; Wendisch, M.; Helas, G. *Beitr. Phys. Atmos.* **1997**, *70*, 249-263.

3. Satheesh, S. K. *Geophys. Res. Lett.* **2002**, *29*, doi:10.1029/2002GL015334.
4. Satheesh, S. K. *Curr. Sci.* **2002**, *82*, 310-316.
5. Hansen, J.; Sato, M.; Ruedy, R. *J. Geophys. Res.* **1997**, *102*, 6831-6864.
6. Lohmann, U.; Kärcher, B. *J. Geophys. Res.* **2002**, *107*, doi:10.1029/2001JD000767.
7. Lacis, A. A.; Mishchenko, M. I. In *Aerosol forcing of climate, report of the Dahlem workshop on aerosol forcing of climate, Berlin 1994*; Charlson, R. J., Heintzenberg, J., Eds.; John Wiley and Sons Ltd.: Chichester, New York, 1995.
8. Schwartz, S. E., et al. In *Aerosol forcing of climate, report of the Dahlem workshop on aerosol forcing of climate, Berlin 1994*; Charlson, R. J., Heintzenberg, J., Eds.; John Wiley and Sons Ltd.: Chichester, New York, 1995.
9. Chou, M.-D.; Chan, P.-K.; Wang, M. *J. Atmos. Sci.* **2002**, *59*, 748-767.
10. Mishchenko, M. I.; Travis, L. D. *Journal of Quantitative Spectroscopy & Radiative Transfer* **1994**, *51*, 759-778.
11. Mishchenko, M. I.; Travis, L. D. *Appl. Opt.* **1994**, *33*, 7206-7225.
12. Hobbs, P. V. *Aerosol-cloud-climate interactions* Academic Press: San Diego, CA, 1993, pp 235.
13. Textor, C., et al. *Atmos. Chem. Phys.* **2006**, *6*, 1777-1813.
14. Kinne, S., et al. *Atmos. Chem. Phys.* **2006**, *6*, 1815-1834.
15. *Climate Change 2007: The Physical Science Basis*; Solomon, S.; Qin, D.; Manning, M.; Chen, Z.; Marquis, M.; Averyt, K. B.; Tignor, M.; Miller, H. L., Eds.; Cambridge Univ. Press: New York, NY, 2007.
16. Lohmann, U.; Feichter, J. *Atmos. Chem. Phys.* **2005**, *5*, 715-737.
17. Takemura, T.; Nakajima, T.; Dubovik, O.; Holben, B. N.; Kinne, S. *J. Clim.* **2002**, *15*, 333-352.
18. Jacobson, M. Z. *J. Geophys. Res.* **2001**, *106*, 1551-1568.
19. Lesins, G.; Lohmann, U. *J. Atmos. Sci.* **2003**, *60*, 2747-2764.
20. Haywood, J. M.; Ramaswamy, V.; Soden, B. J. *Science* **1999**, *283*, 1299-1303.
21. Ayash, T.; Gong, S. L.; Jia, C. Q. *J. Clim.* **2007**, accepted.
22. Woodward, S. *J. Geophys. Res.* **2001**, *106*, 18,155-18,166.
23. Sokolik, I. N.; Andronova, A.; Johnson, T. C. *Atmos. Environ.* **1993**, *27*, 2495-2502.
24. Dobbie, S.; Li, J.; Harvey, R.; Chylek, P. *Atmos. Res.* **2003**, *65*, 211-233.
25. Grini, A.; Myhre, G.; Sundet, J. K.; Isaksen, I. S. A. *J. Clim.* **2002**, *15*, 1717-1730.
26. Tegen, I. *Quat. Sci. Rev.* **2003**, *22*, 1821–1834.
27. Miller, R. L.; Perlwitz, J.; Tegen, I. *J. Geophys. Res.* **2004**, *109*, doi:10.1029/2004JD004912.
28. Perlwitz, J.; Tegen, I.; Miller, R. L. *J. Geophys. Res.* **2001**, *106*, 18,167-18,192.

29. Ayash, T. Ph.D. thesis, University of Toronto, Toronto, Canada, 2007.
30. Liepert, B. G.; Feichter, J.; Lohmann, U.; Roeckner, E. *Geophys. Res. Lett.* **2004**, *31*, doi:10.1029/2003GL019060.
31. Takemura, T.; Nozawa, T.; Emori, S.; Nakajima, T. Y.; Nakajima, T. *J. Geophys. Res.* **2005**, *110*, doi:10.1029/2004JD005029.
32. Feichter, J.; Roeckner, E.; Lohmann, U.; Liepert, B. *J. Clim.* **2004**, *17*, 2384-2398.
33. Menon, S.; Hansen, J.; Nazarenko, L.; Luo, Y. *Science* **2002**, *297*, 2250-2253.
34. Ramanathan, V., et al. *J. Geophys. Res.* **2001**, *106*, 28,371-28,398.
35. Chung, C. E.; Ramanathan, V.; Kiehl, J. T. *J. Clim.* **2002**, *15*, 2462-2476.
36. (36) Ramanathan, V.; Crutzen, P. J.; Kiehl, J. T.; Rosenfeld, D. *Science (Wash.)* **2001**, *294*, 2119-2124.
37. Chung, C. E.; Zhang, G. J. *J. Geophys. Res.* **2004**, *109*, doi:10.1029/2004JD004726.
38. Williams, K. D.; Jones, A.; Roberts, D. L.; Senior, C. A.; Woodage, M. *J. Clim. Dyn.* **2001**, *17*, 845-856.
39. Rotstayn, L. D.; Lohmann, U. *J. Clim.* **2002**, *15*, 2103–2116.
40. Roberts, D. L.; Jones, A. *J. Geophys. Res.* **2004**, *109*, doi:10.1029/2004JD004676.
41. Chung, C. E.; Ramanathan, V. *J. Clim.* **2003**, *16*, 1791-1806.
42. Coakley, J. A.; Cess, R. D. *J. Atmos. Sci.* **1985**, *42*, 1677-1692.
43. Miller, R. L.; Tegen, I.; Perlwitz, J. *J. Geophys. Res.* **2004**, *109*, doi:10.1029/2003JD004085.
44. Kohfeld, K.; Harrison, S. P. *Earth-Sci. Rev.* **2001**, *54*, 81-114.
45. Bush, A. B. G.; Philander, S. G. H. *J. Geophys. Res.* **1999**, *104*, 24,509-24,525.

Chapter 11

Environmental Effects to Residential New Orleans following Hurricane Katrina: Indoor Sediment as Well as Vapor-Phase and Aerosolized Contaminants

Nicholas A. Ashley, Kalliat T. Valsaraj, and Louis J. Thibodeaux

Cain Department of Chemical Engineering, Louisiana State University, Baton Rouge, LA 70803

The flooding of New Orleans following Hurricane Katrina resulted in one of the most complex and widespread urban environmental disasters encountered in modern times. Hundreds of thousands of homes and businesses were destroyed as floodwaters inundated the city, bringing with them sediment containing metal and organic contaminants from Lake Pontchartrain. The sediment and associated contaminants which found their way inside homes represent a special problem. Indoor pollutants are found to be more highly concentrated than their external counterparts, in some cases by as much as a factor of ten. Interior pollutants also act as a direct exposure source to New Orleans' residents, first responders, and recovery personnel. Because of the hot, damp conditions experienced in many of the homes, heavy mold growth developed along the walls, and subsequently emitted very high concentrations of mold spores into the vapor space inside the homes. These aerosolized mold spores can serve as a sink for vapor-phase pollutants, thereby increasing the total inhalation exposure of residents and other workers to toxic materials.

Introduction

On August 29, 2005, the United States experienced the worst natural disaster in American history when Hurricane Katrina made landfall on the Louisiana and Mississippi Gulf coasts. Katrina's gale force winds, heavy rain fall, and massive storm surge overwhelmed many of the protective levees around the city of New Orleans, resulting in floodwaters from Lake Pontchartrain and associated drainage canals inundating residential portions of the city. After the storm passed, floodwaters sat stagnant in the city for nearly two weeks until electricity could be restored, and drainage pumps began to pump the floodwaters back into the Lake. As the floodwaters receded and the damage began to be assessed, New Orleans was faced with an epic problem, with many more environmental questions than answers. Hundreds of thousands of homes and businesses were flood-damaged, with water-lines on homes and other structures ranging from a few feet to complete submersion of the home. No information was available concerning what kind of contaminants had been brought in by the floodwaters, and whether or not they were toxic to the people who may have been wading through them in an attempt to be rescued. As the U.S. EPA, other government agencies, and research groups began to assess the situation, efforts were mainly focused on the floodwaters themselves and on sediment deposits in exterior areas, such as on roadways, in playgrounds, schools, and yards. Contaminants which may have been deposited inside flooded homes were largely ignored, yet represent a key piece of the environmental puzzle in post-Katrina New Orleans, because of their ability to serve as a direct exposure to first responders and returning residents. A complete assessment of in-home contamination must include the four primary phases where in-home pollutants may be found: sediment, vapor-phase, aerosolized mold spores, and mold films growing on walls and other interior surfaces. This article describes such an analysis, as well as suggests key focus areas for future research.

Residential Flooding and Particle Winnowing

As floodwaters from Lake Pontchartrain and drainage canals poured into the city, they carried with them suspended sediments from the Lake, churned up by Katrina's high winds *(1)*. Pontchartrain sediments are known to have been previously contaminated with many different types of chemicals, including alkanes from refined petroleum products, pesticides and metals from agricultural runoff, lead from paint and vehicular emissions, and many other types of pollutants *(2-9)*. As floodwaters flowed across levee breaches, through yards, playgrounds, and neighborhoods, any metal and organic soil pollutants could also be accumulated into the floodwaters and mix with the suspended sediments.

As floodwaters enter a home, two processes must occur: first, the water must dramatically slow in velocity, reducing turbulence; second, the water must seep in through small cracks around doors and windows. Both processes result in a loss of energy for the floodwaters, and cause larger sediment particles to fall out on the outside of the home. This implies that the only material reaching the inside of the home are the finest particles, comprised mostly of silts and clays, which then settle out on the floors of the home once the floodwaters stagnate and remain behind once the water has been pumped back into Lake Pontchartrain *(10)*. This process is illustrated in Figure 1.

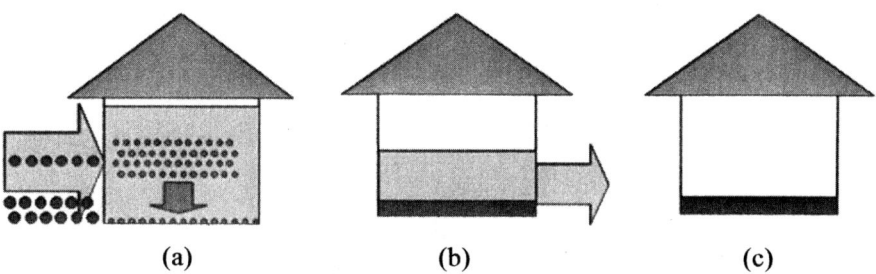

Figure 1. Particle winnowing/settling process from Hurricane Katrina floodwaters. (Reproduced from reference (11). Copyright 2008 American Society of Civil Engineers.)

As many homes remained closed-up for weeks to months after the storm, pollutants that entered the homes with floodwater suspended sediments distributed themselves among several different phases. The vapor space inside the homes is large, and can serve as a significant sink for volatile species, as well as some semi-volatile pollutants that can partition from the sediment. The warm, humid conditions encountered during August and September in south Louisiana also led to the growth of mold on the damp walls, upholstery, and other surfaces. The mold films emit mold spores into the air, which remain suspended in an organic aerosol phase inside the vapor space of the home as well, and can also serve as a sink/partitioning medium for volatile and semi-volatile species. Thus, there are a total of four different phases where in-home pollutants may reside: sediment, vapor-phase, mold films, and aerosolized mold spores. In order to determine the extent of in-home contamination of Katrina-flooded homes, as well as the distribution of contaminants across all four phases, a comprehensive experimental and modeling effort was undertaken.

Experimental Methods and Model Development

Sediment Experimental Analysis

Sediment was collected from inside two flood-damaged homes, and analyzed for metal and organic pollutants. Metals were extracted by microwave-assisted acid digestion, and analyzed by inductively coupled plasma mass spectroscopy (ICP-MS), based on EPA methods 3051 and 6020. Organic contaminants were extracted ultrasonically in acetone/hexane, and analyzed by gas chromatography-mass spectroscopy (GC-MS) based on EPA methods 3550B and 8270C. Further details on both analyses are available elsewhere *(10)*.

Vapor-Phase Modeling

The distribution of contaminants between the sediment and vapor-phase is controlled by properties such as the sediment moisture content and specific surface area, as well as the species vapor pressure and molecular weight. In order to ascertain the maximum possible concentrations in the vapor-phase, a worst-case scenario modeling approach was employed. The worst-case approach assumes the sediment is completely saturated with water, thereby driving the maximum possible amount of volatile and semi-volatile materials into the vapor phase. This thermodynamic equilibrium model is useful because it sets an upper bound on the gas phase concentrations of the species of interest. Sediment-air partitioning be described by the Henry's law constant and sediment-water partition coefficient. The sediment-air partition coefficient, K_{SG} (atm kg/mg) is given by equation 1:

$$K_{SG} = \frac{H}{K_{SW}} \qquad (1)$$

where K_{SW} (L/mg) is the sediment-water partition coefficient, given by equation 2:

$$K_{SW} = f_{oc} K_{oc} \qquad (2)$$

f_{oc} is the sediment's fraction organic carbon, and K_{oc} (L/kg) is the species organic carbon partition coefficient.

Physicochemical properties for all species have been obtained from the literature and standard references. These parameters, along with experimental data on the sediment composition, enable the calculation of vapor-phase concentrations of all compounds of interest.

In computing the vapor-phase concentrations of individual species, two separate equations are possible, depending on the comparison of the measured value of the sediment loading (w_{AS}) to the calculated critical sediment loading (w_{AC}). The critical sediment loading is the concentration in the sediment that would be in equilibrium with the maximum possible vapor-phase concentration, given by the species' saturated vapor pressure. A measured value of the sediment loading which exceeds the calculated critical value implies that a large enough quantity of pollutant is present such that the vapor-phase concentration would be in equilibrium with the pure condensed phase material.

The equations for calculating the concentrations of individual pollutants across multiple phases will be shown here for completeness, but will be presented without derivation. Full development of the models and derivation of the equations are available elsewhere *(12)*. Final equations for computing the vapor-phase concentrations, as well as mold phase concentrations are shown in Table I.

Aerosol Modeling

Partitioning of vapor-phase pollutants into an organic-rich aerosol, such as mold spores, may represent a significant sink for certain classes of contaminants. An aerosolized mold phase is of utmost concern because of its ability to be inhaled by returning residents, potentially increasing the mass exposure to certain toxic chemicals.

We begin the modeling analysis by considering octanol to be representative of the organic and biological fraction of the mold spore. The octanol-air partition coefficient, K_{OA}, can then be used to describe partitioning of organic pollutants between the particle and vapor phases. A mold-gas phase partition coefficient (K_{MG}) is thus defined based on K_{OA}, and the spore's fraction organic carbon, f_{oc}:

$$K_{MG} = f_{oc} K_{OA} \qquad (3)$$

K_{MG} (μg/m^3 mold / μg/m^3 air) can then be used to predict particle mass loadings, using predicted vapor-phase concentration results from the previous model. Model parameters, including the fraction organic carbon (f_{oc}) and spore wet density, are chosen so as to provide a conservative estimate of spore pollutant loadings. The reader is reminded that model predictions are not intended as absolute measures of pollutant mass loadings, but rather should serve as a first approximation towards assessing the contaminant distribution across multiple phases inside flooded homes on a broad scale. Equations for calculating the mold spore loadings were shown in Table I.

Table I. Equations for computing multi-phase concentrations of in-home pollutants.

	$w_{AS} < w_{AC}$	$w_{AS} > w_{AC}$
Vapor Phase	$\dfrac{w_{AS} K_{SG} MW_A}{RT}$	$\dfrac{P_A^{sat} \cdot MW_A}{RT}$
Mold Film	$\dfrac{w_{AS} K_{SG} K_{MG} V_{film} MW_A}{RT m_{film}}$	$\dfrac{P_A^{sat} MW_A K_{MG}}{RT \rho_{mold}}$
Mold Spores	$\dfrac{w_{AS} K_{SG} K_{MG} C_{Mvap} \cdot MW_A}{RT \rho_{mold}}$	$\dfrac{P_A^{sat} MW_A C_{Mvap} K_{MG}}{RT \rho_{mold}}$

NOTE: MW_A = molecular weight of species A; K_{MG} = mold-gas phase partition coefficient (defined below); m_{film} & V_{film} = mass and volume of mold film; C_{Mvap} = concentration of mold aerosol in the vapor phase; ρ_{mold} = corrected wet density of mold
Units are: μg/m³ (vapor); mg/kg (mold film); ng/m³ (mold spores)
SOURCE: Reproduced with permission from reference 12. Copyright 2008 American Society of Civil Engineers.

Film Modeling

In addition to the aerosolized mold spores, vapor-phase contaminants can also partition between the gas-phase and the mold films growing on walls and other interior surfaces. Chemically, we assume that the mold spores and mold film are comprised of the same material (equivalent to assuming that the film is a collection of mold spores, as has been previously shown (13)). Thus, the same modeling approach that was taken for the aerosol phase can be applied here, the only difference lies in the representative volumes for the particle and film phases. Based on direct observation and on numerous images of damaged flooded homes taken after the storm (14), we conservatively estimate the interior mold surface coverage to be 10 m². Certainly there are structures with much greater mold surface coverage than our estimate. Film thicknesses were estimated by considering typical thicknesses of urban organic films, which have been extensively studied by Diamond and co-workers (15-17). They report film thicknesses ranging from 29-250 nm, with 70 nm films being an average thickness (16, 17). We thus consider a mold film of approximately 7×10^{-8} m³ in volume for this analysis. Equations for calculating the mold film loadings were shown in Table I.

Currently, little is known regarding volatile or semi-volatile organic chemical partitioning among either mold film or aerosolized mold spores. The equilibrium approach employed here is a first approximation technique, designed to probe the affinity which certain classes of compounds may have for the mold phases. Clearly, more detailed experimental analysis is needed in this area, and such data can be used to form the basis of more refined equilibrium distribution models.

Overall Mass Distribution

Based on the results of the vapor, aerosol, and film models, as well as the experimental sediment analysis, a global mass distribution of contaminants across all phases inside the home can be obtained. A breakdown into phases by contaminant mass can serve as a useful tool for researchers, first responders, returning residents, and recovery workers in the region, by highlighting which classes of pollutants may be most prevalent in a particular phase.

The mass distribution model is constructed by a Lavosier species mass balance approach, which accounts for the pollutant mass that entered the home with the sediment, as well as the final equilibrium distribution of the pollutant across all four indoor phases (sediment, vapor, aerosol, film). The ratio of the species mass in each phase to the total mass of the pollutant gives the relative distribution of each species/pollutant class across the multiple indoor phases. Further details on the development of all the models described herein are available elsewhere *(11, 12)*.

Results and Discussion

Sediment Geochemistry

A particle size analysis was conducted on in-home sediment samples to determine the sediment geochemistry and overall particle size distribution. Results are shown in Figure 2.

Results of the particle size analysis show that the in-home sediments are overwhelmingly dominated by silts and clays, with at least 92% of particles residing in this regime *(10)*. In terms of particle diameter, silts are defined as particles having diameters between 4 and 63 μm, with clays having diameters below 4 μm. Comparison with Lake Pontchartrain sediment seems to support the particle winnowing process described previously; Pontchartrain sediments

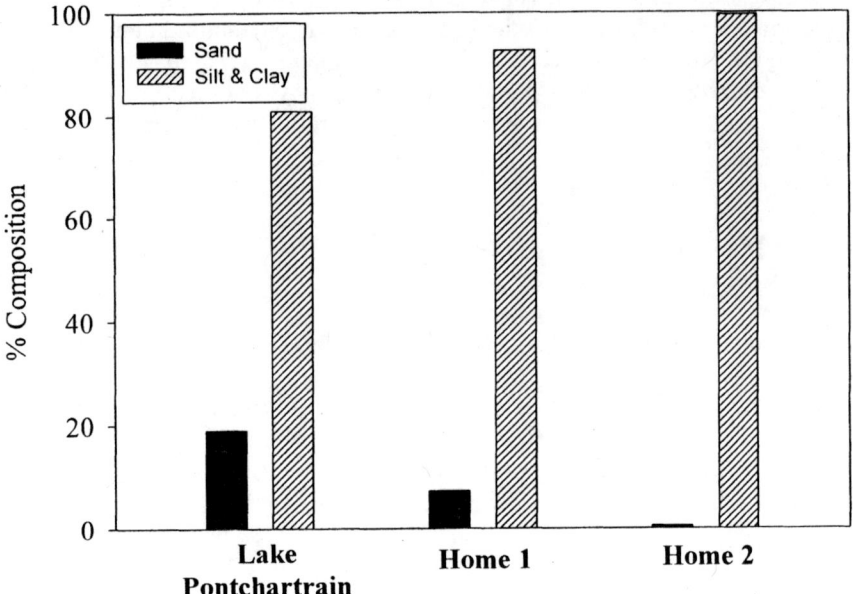

Figure 2. Particle size distribution of Lake Pontchartrain and in-home sediments. Data from references (5, 10).

are nearly 20% sand (particles having diameter greater than 63 μm), whereas for the in-home sediments the maximum sand content is between 2 and 8 %.

Sediment geochemistry is important because the types of particles present in a sediment have great influence over both the type and amounts of contaminants which may be found. Silt and clay particles have high specific surface areas, and large quantities of adsorption sites on the mineral surface. These adsorption sites can serve as hosts for metal and organic pollutants, and it has been shown by many researchers that the degree of contamination of a sediment increases with increasing silt and clay content *(18-20)*. Sediment organic matter also plays a role in sequestering pollutants from floodwaters into sediment material. In-home sediments were found to have organic carbon contents ranging from 7-9% *(10)*; volatile and semi-volatile pollutants may partition into the sediment organic matter, in addition to adsorbing to active sites on the mineral surfaces.

The sediment geochemistry analysis has important consequences for the overall post-Katrina environmental analysis. Because the in-home sediments are overwhelmingly dominated by fine particles (silts and clays), whereas coarser particles (sand) are expected to be winnowed out on the exterior, the potential exists for in-home sediments to be more contaminated than their exterior counterparts. This would have important implications for exposure to returning residents and recovery workers. Sediment pollutant concentrations will be explored more fully in the following section.

In-Home Sediment Contaminants

Metals

Average sediment metal loadings for the two flooded homes are given in Table II. Of the twenty metals analyzed, four were detected in high concentrations, including arsenic, cadmium, lead, and vanadium.

Table II. Average metal concentrations from sediment inside two flooded homes.

	Home 1	*Home 2*
Arsenic	66	194
Cadmium	381	104
Lead	274	1500
Vanadium	64	45

NOTE: Units are mg/kg dry sediment
SOURCE: Data from reference *(10)*.

Arsenic, cadmium and lead sediment concentrations exceed their EPA soil screening standards by 1.5 – 10 times. Compared with an exterior sediment sample taken from between the two homes, arsenic concentrations are two times higher, cadmium concentrations nearly ten times higher, and lead concentrations four times higher than the exterior sample *(21)*. One of the primary reasons for the elevation of the in-home samples to the exterior ones must be attributed to the sediment particle size geochemistry as described earlier.

Comparison with New Orleans soil background concentrations from pre-Katrina studies shows that for arsenic, cadmium, and vanadium, the in-home, flood-deposited sediments are more highly contaminated compared to previous levels of exposure. Average indoor arsenic and vanadium concentrations are nearly four times greater, and cadmium nearly fifty times greater than pre-Katrina outdoor soil background concentrations *(10, 22)*. Lead concentrations were in the range of previously measured soil concentrations, due largely to the fact that New Orleans soils historically contain very large quantities of lead *(7, 8, 23, 24)*.

High loadings of arsenic, cadmium, and lead may pose a direct risk to returning residents, first responders, and recovery workers. Persons may be exposed to these contaminants while trying to recover items from damaged homes, or while trying to clean up and renovate the homes. Based on the above data, it is also evident that "safety" warnings and bulletins based solely on measured exterior sediment concentrations may under-predict the real exposure of toxic metals to returning residents.

Organics

Selected results of the sediment organics analysis, along with modeling results for the vapor-phase, aerosolized mold spores, and mold film are shown in Table III.

Types of compounds detected include the alkanes from motor vehicle fuel, aldehydes, esters from plasticizers, organic acids, pesticides from agricultural runoff and termite control, and PAHs and volatiles from fuels. In particular, the pesticides and PAH fluoranthene are detected in high concentrations. The concentrations of chlordane and dieldrin, two common pesticides, are both 100 times greater than the EPA soil regulatory standard *(11)*. The concentration of the PAH fluoranthene is nearly 20 times that detected in an outdoor sample located between the two flooded homes *(21)*. These results, combined with the comparison of indoor and outdoor sediment metal concentrations, lends further credence to the theory that in-home sediment contaminants may be present at higher loadings as a result of the particle winnowing process. The high loadings of pesticides relative to their EPA screening standard, along with the large increase in fluoranthene concentration compared to the exterior sample again reiterate the fact that advisories to returning residents based solely on exterior sampling may under-predict the actual exposure to certain hazardous pollutants.

Predicted Vapor-Phase Contaminants

Predicted results for the indoor vapor-phase concentrations were also shown in Table III. The results show that some of the aldehydes, alkanes, esters, pesticides, and volatiles may be present in significant concentrations in the vapor-phase.

The volatile trimethyl benzenes have very high vapor pressures, and their appearance in significant quantities in the vapor space inside the home is expected. Organic acids are also common vapor-phase pollutants *(25)*. Potentially high concentrations of the pesticide chlordane may represent a particular area of concern for returning residents, as long-term inhalation exposure can lead to chronic health conditions *(26)*.

Table III. Measured and predicted concentrations for selected pollutants across multiple indoor phases.

Compound	Measured Sediment (mg/kg wet)		Vapor Phase ($\mu g/m^3$ air)		Mold Aerosol (ng/m^3 air)		Mold Film (mg/kg film)	
	Home 1	Home 2	Home 1	Home 2	Home 1	Home 2	Home 1	Home 2
Chlordane	280	BDL	76.4	BDL	267	BDL	32,200	BDL
Dieldrin	3.0	BDL	1.4	BDL	0.8	BDL	93.9	BDL
Fluoranthene	35.6	46.5	6.5	6.9	20.7	21.9	2490	2650
Heptadecane	3.3	2.3	1320	1290	1.7	3.7	207	452
Nonanal	BDL	7.8	BDL	31,300	BDL	8.2	BDL	985
Oleic Acid	338	697	69.5	67.9	--	--	--	--
Phthalic anhydride	BDL	292	BDL	260	BDL	65.1	BDL	7860
Tetracosane	105	37.1	1.6	1.5	0.8	2.2	93.9	269
Trimethyl benzenes	3.5	BDL	15,400	BDL	1.1	BDL	129	BDL

NOTE: BDL = below detection limit in sediments; mold model unable to be processed for organic acids
SOURCE: Data from references *(10, 12)*

Sediment moisture content plays a key role in the distribution of pollutants between the sediment and vapor phases, because water competes with organic species for adsorption sites on clay mineral surfaces. A damp or saturated sediment can force volatile and semi-volatile organic molecules into the vapor phase, where they may function as an inhalation exposure source to persons entering the flood-damaged homes *(27, 28)*. Therefore, the worst-case wet sediment modeling approach employed herein is useful because it predicts a maximum possible indoor vapor phase concentration, and sets an upper limit on the concentrations to which residents and recovery workers may have been exposed to volatile and semi-volatile pollutants in the gas phase.

Predicted Aerosolized Mold Spore Contaminants

Results of the modeling work for aerosolized mold spores, also shown in Table III, reveal that some esters, pesticides, and the PAH fluoranthene may partition into the aerosol phase in appreciable concentrations. These results are significant because they demonstrate that the aerosolized mold spores, present in nearly every flooded home in New Orleans following Hurricane Katrina, can act as a partitioning medium for certain classes of pollutants from the vapor phase.

Mold spore concentrations inside flooded homes in excess of 500,000 spores/m^3 have been measured *(29)*. It is clear that at these extremely high concentrations, inhalation of spores by returning residents and recovery workers is unavoidable. The N-95 respirators recommended by many agencies have been shown to have little effect against spores *(30)*. If the spores can also contain organic pollutants, then the total inhalation exposure to certain classes of contaminants will be the sum of the vapor-phase and aerosol-associated masses. Exposure assessments which only accounted for vapor-phase pollutants may then under-predict the real amount of exposure.

The high indoor mold spore concentrations discussed above will have the effect of creating a highly concentrated aerosol phase inside the flooded homes. Because the mold spores are comprised mainly of water, organic, and biological components *(31)*, it is completely plausible to believe that mold can act as a partitioning medium for volatile and semi-volatile species in the gas phase. The idea of mold spores acting as a sink for organic pollutants in flooded homes has not previously been considered, yet represents a key piece of the total environmental picture in post-Katrina New Orleans. Modeling results indicate that many of the alkanes may also be present in considerable amounts in the aerosolized mold spores *(12)*. This is somewhat surprising given the low vapor-phase concentrations of the alkanes. However, many of the organic species comprising the mold spores, such as lignin and other long-chain hydrocarbons, are chemically very similar to the alkanes, and as such they may have a special affinity for the mold. The pesticide chlordane, fluoranthene, a carcinogenic PAH, and the volatile benzene derivatives also show potentially high loadings in the aerosol phase. These results are especially important because chlordane and the trimethyl benzenes also showed the potential for high vapor-phase concentrations; therefore their presence in the aerosol phase could lead to a very high total inhalation exposure of these compounds. Overall, modeling results for the aerosolized mold spores suggest that their presence should not be ignored in understanding the total chemodynamic behavior of indoor contaminants from flooded homes. For specific compound classes, the mold spores may also contribute substantially to the total inhalation exposure to toxic materials.

Predicted Mold Film Contaminants

Compared to the aerosolized mold spores where the primary means of exposure will be through inhalation, mold films growing on walls and other interior surfaces represent exposure concerns through dermal contact, as well as present issues to be considered in the disposal of housing debris. Predicted mold film concentrations were also provided in Table III.

Modeling results indicate that pesticides and esters may be the most significant pollutants in this phase. When destroyed homes are torn down or gutted, contaminated debris including drywall, upholstery, furniture, and other items that may contain a mold film are often disposed of in landfills or other waste disposal facilities. Pesticides and esters, which may be present in high concentrations in the mold film, are thus also introduced to the landfills or incinerators. Careful monitoring of effluent streams and by-products of these facilities should be conducted to ensure that no hazardous materials are leaving the site.

Overall Mass Distribution

The summary results of the overall mass distribution model are shown in Table IV. The mass distribution model shows which classes of pollutants may be the most important in each specific phase, giving a "snapshot" view of the distribution of contaminants across multiple phases inside New Orleans flooded homes.

Results show that the sediment is dominated by alkanes and organic acids, the vapor-phase by aldehydes, volatiles, and organic acids, and the mold phases (both film and aerosolized spores) by pesticides, esters, and PAH. Several important exposure conclusions can be drawn based on the mass distribution. First, organic acids are prevalent in both the sediment and vapor-phases, and thus may be significant in terms of both dermal and inhalation exposure. Second, because pesticides and esters are large contributors to both the vapor and mold phases, their additive contribution to respirable vapors and aerosolized mold spores may be very significant, and neglecting the aerosolized mold phase may under-predict the total exposure to these chemicals.

Conclusions and Future Work

Through experimental analysis and modeling efforts, the total pollutant distribution inside flooded homes and thus the overall exposure scenario for returning residents, first responders, and recovery workers can be obtained. This work, however, is far from complete. The idea of aerosolized mold spores as a partitioning and transport medium for organic pollutants is novel, and a broad experimental effort is needed in order to gain understanding of this complex situation, as well as to enable the development and refinement of models to predict pollutant behavior in the presence of high concentrations of mold aerosol.

Table IV. Overall mass distribution of in-home contaminants

Pollutant Class	Sediment		Vapor		Mold	
	Home 1	Home 2	Home 1	Home 2	Home 1	Home 2
Aldehydes	tr	tr	tr	78 %	tr	7 %
Alkanes	80 %	22 %	3 %	4 %	1 %	13 %
Esters	3 %	3 %	tr	tr	86 %	59%
Organic Acids	10 %	73 %	tr	18 %	tr	tr
Pesticides	5 %	tr	tr	tr	12 %	tr
PAH	2 %	1 %	tr	tr	1 %	20 %
Volatiles	tr	tr	96 %	tr	tr	tr

NOTE: tr = trace (<1 %)
SOURCE: Data from reference *(12)*

Finally, it is our hope that this work will shed some light on a key area of environmental concern that was widely overlooked in the aftermath of Hurricane Katrina, the state of the hundreds of thousands of flooded homes with regards to hazardous chemical pollutants. Initial reports and analysis after the storm were focused on floodwaters and on exterior deposited sediment. We have demonstrated that contaminant concentrations and distributions inside the flooded homes may differ from those on the exterior, and also that in-home pollutants pose a more direct exposure risk to many people returning to and working in the region. It is our hope that in future urban flood situations, where homes or other isolated structures have the potential to become contaminated, that more careful attention will be paid to pollutants inside these structures, and not only on exterior contaminants.

Acknowledgements

The authors are grateful for two homeowners, Ms. Deweylene B. Schneider and Ms. Leatrice J. Burr, who allowed them to sample inside their destroyed homes. We also thank Mr. Melvin N. Schneider, III, for his assistance with the sampling. Nick Ashley acknowledges support of a Donald W. Clayton Graduate Fellowship through the LSU College of Engineering. This work was funded by the Cain Department of Chemical Engineering at Louisiana State University.

References

1. Bianchi, T. S.; Argyrou, M. E., *Estuar. Coast. Shelf. S.* **1997**, *45*, 557-569.
2. Overton, E. B.; Schurtz, M. H.; St. Pe, K. M.; Byrne, C., *ACS Symp. Ser.* **1986**, *305*, 247-270.
3. McFall, J. A.; Antoine, S. R.; DeLeon, I. R., *Chemosphere* **1985**, *14*, 1561-1569.
4. McFall, J. A.; Antoine, S. R.; DeLeon, I. R., *Chemosphere* **1985**, *14*, 1253-1265.
5. Flowers, G. C.; Isphording, W. C., *T. Gulf Coast Ass. Geo.* **1990**, *40*, 237-250.
6. DeLaune, R. D.; Gambrell, R. P., *J. Environ. Sci. Heal. A.* **1996**, *A31*, 2349-2362.
7. Mielke, H. W., *Environ. Geochem. Health* **1994**, *16*, 123-128.
8. Mielke, H. W.; Wang, G.; Gonzales, C. R.; Le, B.; Quach, V. N.; Mielke, P. W., *Sci. Total Environ.* **2001**, *281*, 217-227.
9. Manheim, F. T.; Flowers, G. C.; McIntire, A. G.; Marot, M.; Holmes, C., *T. Gulf Coast Ass. Geo.* **1997**, *47*, 337-350.
10. Ashley, N. A.; Valsaraj, K. T.; Thibodeaux, L. J., *Chemosphere* **2008**, *70*, 833-840.
11. Ashley, N. A.; Valsaraj, K. T.; Thibodeaux, L. J., In *Geocongress 2008: Geosustainability and Geohazard Mitigation*, Reddy, K. R.; Khire, M. V.; Alshawabkeh, A. N., Eds.; ASCE Geotechnical Special Publications # 178, New Orleans, LA, 2008; Vol. 178, pp 425-432.
12. Ashley, N. A.; Valsaraj, K. T.; Thibodeaux, L. J., *Environmental Engineering Science* **2008**, In Review.
13. Schwab, K. J.; Gibson, K. E.; Williams, D. A. L.; Kulbicki, K. M.; Lo, C. P.; Mihalic, J. N.; Breysse, P. N.; Curriero, F. C.; Geyh, A. S., *Environ. Sci. Technol.* **2007**, *41*, 2401-2406.
14. Polidori, R., *After the Flood*; Steidl: Göttingen, 2006.
15. Law, N. L.; Diamond, M. L., *Chemosphere* **1998**, *36*, 2607-2620.
16. Diamond, M. L.; Gingrich, S. E.; Fertuck, K.; McCarry, B. E.; Stern, G. A.; Billeck, B.; Grift, B.; Brooker, D.; Yager, T. D., *Environ. Sci. Technol.* **2000**, *34*, 2900-2908.
17. Diamond, M. L.; Priemer, D. A.; Law, N. L., *Chemosphere* **2001**, *44*, 1655-1667.
18. Kleineidam, S.; Rugner, H.; Grathwohl, P., *Environ. Toxicol. Chem.* **1999**, *18*, 1673-1678.
19. Talley, J. W.; Ghosh, U.; Tucker, S. G.; Furey, J. S.; Luthy, R. G., *Environ. Sci. Technol.* **2002**, *36*, 477-483.
20. Shor, L. M.; Rockne, K. J.; Taghon, G. L.; Young, L. Y.; Kosson, D. S., *Environ. Sci. Technol.* **2003**, *37*, 1535-1544.

21. *EPA, U. S. FLOODWATER : Sediment Chemical Testing Results for 9548: September 18, 2005.* URL http://oaspub.epa.gov/storetkp/storet_wme_pkg.Station_Sediment_Chem_Results?p_station_id=9548&p_org_id=KATRINA6&p_sample_date=09-18-2005.
22. Presley, S. M.; Rainwater, T. R.; Austin, G. P.; Platt, S. G.; Zak, J. C.; Cobb, G. P.; Marsland, E. J.; Tian, K.; Zhang, B.; Anderson, T. A.; Cox, S. B.; Abel, M. T.; Leftwich, B. D.; Huddleston, J. R.; Jeter, R. M.; Kendall, R. J., *Environ. Sci. Technol.* **2006**, *40*, 468-474.
23. Mielke, H. W.; Wang, G.; Gonzales, C. R.; Powell, E. T.; Le, B.; Quach, V. N., *Environ. Toxicol. Pharmacol.* **2004**, *18*, 243-247.
24. Wang, G.; Mielke, H. W.; Quach, V.; Gonzales, C.; Zhang, Q., *Soil Sediment Contam.* **2004**, *13*, 313-327.
25. Hays, M. D.; Stockburger, L.; Lee, J. D.; Vette, A. F.; Swartz, E. C., In *Urban Aerosols and Their Impacts: Lessons Learned from the World Trade Center Tragedy*; Gaffney, J. S.; Marley, N. A., Eds.; ACS Symposium Series # 919. Oxford University Press: Washington, D.C., 2006; pp 164-188
26. Hoppin, J. A.; Valcin, M.; Henneberger, P. K.; Kullman, G. J.; Umbach, D. M.; London, S. J.; Alavanja, M. C. R.; Sandler, D. P., *American Journal of Industrial Medicine* **2007**, *50*, 969-979.
27. Valsaraj, K. T.; Thibodeaux, L. J., *J. Hazard. Mater.* **1988**, *19*, 79-99.
28. Ravikrishna, R.; Valsaraj, K. T.; Yost, S.; Price, C. B.; Brannon, J. M., *J. Hazard. Mater.* **1998**, *60*, 89-104.
29. Solomon, G. M.; Hjelmroos-Koski, M.; Rotkin-Ellman, M.; Hammond, S. K., *Environ. Health Perspect.* **2006**, *114*, 1381-1386.
30. Chew, G. L.; Wilson, J.; Rabito, F. A.; Grimsley, F.; Iqbal, S.; Reponen, T.; Muilenberg, M. L.; Thorne, P. S.; Dearborn, D. G.; Morley, R. L., *Environ. Health Perspect.* **2006**, *114*, 1883-1889.
31. Porges, N., *Bot. Gaz.* **1932**, *94*, 197-205.

Indexes

Author Index

Ashley, Nicholas A., 167
Ayash, Tarek, 149
Baer, Tomas, 13
Barbier, C., 79
Bateman, Adam P., 91
Chen, Jing, 127
Chunram, Narongpan, 31
Deming, Richard L., 31
Donaldson, D. J., 79
Ehrenhauser, Franz S., 127
Garland, Eva R., 13
Gomez, Anthony L., 91
Gong, Sunling, 149
Handley, S. R., 79
Hao, Jiming, 111
Jia, Charles Q., 149
Kahan, T. F., 79
Kamens, Richard M., 31
Kang, Ying, 41
Kommalapati, Raghava R., 1

Kwamena, N.-O. A., 79
Li, Junhua, 111
Lu, Zifeng, 111
Mang, Stephen A., 91
Nizkorodov, Sergey A., 91
Pan, Xiang, 91
Park, Jiho, 91
Rontu, Nabilah, 65
Rosen, Elias P., 13
Takekawa, Hideto, 111
Thibodeaux, Louis J., 167
Underwood, Joelle S., 91
Vaida, Veronica, 65
Valsaraj, Kalliat T., 1, 127, 167
Vinitketkumnuen, Usanaee, 31
Walser, Maggie L., 91
Wornat, Mary J., 127
Wu, Zucheng, 41
Xing, Jia-Hua, 91

Subject Index

A

Aerosol cooling, climatic effects, 163
Aerosolized contaminants
  modeling, 171–172
  predicted mold spore, 177–178
  See also Hurricane Katrina aftermath
Aerosol particle reactivity
  oleic acid particles and morphology, 27
  See also Oleic acid coatings
Aerosol particles, reactivity in atmosphere, 14
Aerosols
  condensed phase, 4
  cycling and optical properties, 157, 159
  description, 1
  diagram showing radiative mechanisms with cloud effects in relation to, 5f
  direct effect of, 150
  direct radiative forcing effect on climate, 1
  estimates of, indirect forcing, 160t
  human health and visibility, 4
  indirect effects of atmospheric, 150–151
  mass concentrations and particle sizes, 2t
  optics, 153–154
  pollution, 4
  properties of atmospheric, 2t
  radiative transfer, 156–157
  reflectivity of underlying surfaces, 150
  scattering and absorption of solar radiation by, 150
  statistics of, life-cycle parameters, 158t
  varying atmospheric lifetimes, 1–2
  See also Climatic effects of aerosols; Secondary organic aerosol (SOA)
Airborne aerosol particles
  heterogeneous atmospheric chemistry, 79–80
  See also Atmospheric chemistry
Airborne pollutants
  human health, 32
  See also Particulate matter
Air quality
  Northern Thailand, 32
  See also Particulate matter
Air-water interface, equilibrium partitioning, 130–131
Albedo, reflectivity of underlying surfaces, 150
Ammonium sulfate seed aerosols. *See* Secondary organic aerosol (SOA)
Anthracene
  decay kinetics of, in octanol, 85, 87f
  effect of organic film composition on, kinetics, 82t
  heterogeneous reaction rate in organic films with gas-phase ozone, 81t
Anthropogenic forcing, climate for 2005 vs. 1750, 4, 7f
Aromatic compounds
  emission of volatile organic compounds (VOC), 42
  See also Toluene decomposition
Aromatic hydrocarbons/NO_x photooxidation systems
  categorization of experiments with/without seed aerosols, 121t

comparing SOA yields between aromatic hydrocarbons, 116–117
ozone and secondary organic aerosol (SOA) formation, 111, 112
toluene/NO$_x$, 117, 120f, 121t, 122
variations in SOA in survey experiments of toluene, m-xylene and 1,2,4-trimethylbenzene, 118f, 119f
volatile compounds in atmosphere, 112
m-xylene/NO$_x$ and 1,2,4-trimethylbenzene/NO$_x$, 122, 123f
See also Secondary organic aerosol (SOA)
Arsenic, average concentrations from sediment inside two flooded homes, 175t
Atmosphere
implications of (NH$_4$)$_2$SO$_4$ aerosols, 124–125
interactions between aerosols and gases, 4
perfluorinated compounds, 67
radiative mechanisms associated with cloud effects, 5f
volatile organic compounds (VOCs), 112
See also Secondary organic aerosol (SOA)
Atmospheric aerosols
focus of reviews, 4, 6
organization of book, 6, 8–9
properties, 2t
See also Aerosols
Atmospheric Chemical Transport Model (ACTM), studying aerosols, 153–154
Atmospheric chemistry
chemical reactions on urban films, 80–84
decay kinetics of anthracene in octanol, 87f
effect of organic film composition on anthracene kinetics, 82t
experimental studies, 80–82
fluorescence of acridine in organic film before and after exposure to gas-phase nitric acid, 85, 86f
heterogeneous, of airborne aerosol particles, 79–80
heterogeneous reaction rate for PAHs in organic films with gas-phase ozone, 81t
loss kinetics of benzo[a]pyrene in organic film, 83f
modeling studies, 82, 84
MUM–Fate model, 82, 84
nature of urban surface films, 80
photochemistry of compounds associated with urban organic films, 85–87
rate coefficients for heterogeneous benzo[a]pyrene loss vs. gas-phase ozone concentration, 83f
reactive rate of PAHs in urban environment, 82, 84
sunlight and photolysis of organic compounds, 85
Atmospheric lifetimes, aerosols, 1–2
Atmospheric water films
effect of film thickness on phenanthrene photooxidation rates, 137, 138f
equilibrium uptake of gas-phase phenanthrene, 132–134
photochemical reactions of phenanthrene, 135–137
photooxidation of phenanthrene, 131
See also Phenanthrene, gas phase
Atomic force microscopy (AFM)
method, 21
oleic acid on silica surfaces, 25f
polystyrene spin-coated on silica and oleic acid deposited on spin-coated PS, 26f

B

Beijing, China, particulate matter pollution, 112
Benzo[a]pyrene
　heterogeneous reaction rate in organic films with gas-phase ozone, 81t
　loss kinetics in organic film, 83f
　rate coefficients for, loss vs. gas-phase ozone concentration, 83f
3,4-Benzocoumarin
　effect of Suwannee River fulvic acid (SRFA) on formation rate constants, 137, 139f, 140
　observed formation rate constants, 141, 143f
　product of phenanthrene photooxidation, 135–137
　See also Phenanthrene, gas phase
Bioaerosols, cloud condensation nuclei (CCN), 1

C

Cadmium, average concentrations from sediment inside two flooded homes, 175t
Canadian Aerosol Module (CAM), 161
Canadian Climate Center General Circulation Model (CCC GCM), 161
Cancer, Chiang Mai, Thailand, 33–34
Carbonaceous aerosols, aerosol source, 3
Carbonyls, photodissociation pathways, 101, 102f
Characterization, book overview, 6
Chemistry, book overview, 6, 8
Chiang Mai-Lamphun basin
　Northern Thailand, 32, 33f
　See also Particulate matter

China
　aerosol cooling, 163
　particulate matter pollution, 112
Climate, direct radiative forcing effect of aerosols, 1
Climatic effects of aerosols
　AeroCom annual global statistics for mid-visible aerosol optical thickness (AOT), 158t
　AeroCom statistics of aerosol life-cycle parameters, 158t
　aerosol cycling, 153–154
　aerosol cycling and optical properties, 157, 159
　aerosol optics, 154–156
　aerosol size, 154
　Atmospheric Chemical Transport Model (ACTM), 153–154
　climatic elements in modeling, 151, 153
　climatic implications, 162–164
　connections and dependencies among elements of modeling interactions, 152f
　estimates of aerosol indirect forcing published since 2001, 160t
　estimates of TOA, global, annual-mean direct radiative effect by different aerosol species, 161t
　Mie theory, 155–156
　modeling, 151–162
　radiation transfer, 156–157
　radiative effect, 160–162
　radiative forcing, 159–160
　state of art and uncertainties, 157–162
　Twomey effect, 150–151, 159–160
Climatic seasons, air quality in Northern Thailand, 32
Cloud condensation nuclei (CCN), bioaerosols, 1
Cloud effects, radiative mechanisms associated with, 5f
Coatings. See Oleic acid coatings
Cobalt ions

homogeneous catalytic reactions, 56, 57f
See also Toluene decomposition
Contact angle goniometry
  method, 21
  oleic acid on silica, 25f
Contaminants. *See* Hurricane Katrina aftermath
Corona discharge
  apparatus and procedure for corona radical shower, 56, 58, 59f
  corona radical shower, 42
  toluene degradation route by Fe^{2+} catalyst in corona radical shower, 50, 51f
  See also Toluene decomposition
Cytotoxicity, particulate matter, 34, 36–38

D

Decomposition. *See* Toluene decomposition
Dioctyl sebacate, mixed particles of oleic acid and, 17
n-Docosane system, oleic acid and, 17–18

E

Ellipsometry
  method, 21
  oleic acid on silica, 25f
Environmental effects. *See* Hurricane Katrina aftermath
Equilibrium partitioning, air–water interface, 130–131

F

Fenton reagent

catalysis for compound degradation, 43
effect of catalyst concentration on toluene decomposition, 53
enhancement of toluene decomposition, 50–51, 53
See also Toluene decomposition
Film modeling, mold, 172–173
Flooding of New Orleans. *See* Hurricane Katrina aftermath
Fluoranthene, heterogeneous reaction rate in organic films with gas-phase ozone, 81t
9-Fluorenone
  effect of Suwannee River fulvic acid (SRFA) on formation rate constants, 137, 139f, 140
  observed formation rate constants, 141, 143f
  product of phenanthrene photooxidation, 135–137
  See also Phenanthrene, gas phase
Fluorinated organic compounds. *See* Perfluorinated compounds
Fulvic acid, Suwannee River (SRFA)
  effect on phenanthrene photooxidation rates, 137, 139f, 140
  See also Phenanthrene, gas phase

G

Gas phase phenanthrene. *See* Phenanthrene, gas phase
Geochemistry
  sediment, 173–175
  See also Hurricane Katrina aftermath
Global climate, aerosols, 4
Greenhouse gases, warming effect, 4

H

Henry's law constant, measurement, 131
Heptadecanoic acid, mixed particles of oleic acid and, 17
Hexadecanoic acid, mixed particles of oleic acid and, 17
Human health
  aerosols, 4
  airborne pollutants, 32
Hurricane Katrina aftermath
  aerosol modeling, 171–172
  average metal concentrations from sediment inside two flooded homes, 175t
  equations for computing multi-phase concentrations of in-home pollutants, 172t
  experimental methods, 170–173
  film modeling, 172–173
  future work, 179–180
  in-home sediment contaminants, 175–176
  measured and predicted pollutant concentrations for multiple indoor phases, 177t
  metals, 175–176
  model development, 170–173
  mold-gas phase partition coefficient, 171
  natural disaster, 168
  overall mass distribution, 173, 179, 180t
  particle size distribution of Lake Pontchartrain and in-home sediments, 174f
  particle winnowing/settling process, 169f
  predicted aerosolized mold spore contaminants, 177–178
  predicted mold film contaminants, 178–179
  predicted vapor-phase contaminants, 176–177
  residential flooding and particle winnowing, 168–169
  sediment-air partition coefficient, 170
  sediment experimental analysis, 170
  sediment geochemistry, 173–175
  vapor-phase modeling, 170–171

I

India, aerosol cooling, 163
Indian Ocean, climatic effects of aerosols, 163
Indoor sediment. *See* Hurricane Katrina aftermath
Industrial dust, aerosol source, 3
In-home sediments. *See* Hurricane Katrina aftermath
Inorganic particles. *See* Oleic acid coatings
Inorganic seed aerosols
  particulate matter pollution, 112
  See also Secondary organic aerosols (SOA)
Intergovernmental Panel for Climate Change (IPCC)
  aerosol sources, 2–3
  anthropogenic and natural forcing of climate for 2005 vs. 1750, 4, 7f
  radiative forcing due to atmospheric aerosols, 3–4
Iron ions
  homogeneous catalytic reactions, 56, 57f
  See also Toluene decomposition

K

Katrina. *See* Hurricane Katrina aftermath
Kinetics, oleic acid ozonolysis, 21–23

L

Lake Pontchartrain
  floodwater from, 168–169
  particle size distribution, 174f
  sediment geochemistry, 173–175
  See also Hurricane Katrina aftermath
Langmuir trough. See Perfluorinated compounds
Lauric acid
  experimental and calculated footprint of mixture of, with perfluorododecanoic acid (PFDDA), 71t
  mixture with PFDDA, 70–71
  surface pressure-area isotherm, 68, 69f
  See also Perfluorinated compounds
Lead, average concentrations from sediment inside two flooded homes, 175t
Life-cycle parameters, statistics of aerosols, 158t
Lung cancer
  Chiang Mai, Thailand, 33–34
  cytotoxicity studies of particulate matter, 37–38

M

Manganese ions
  homogeneous catalytic reactions, 56, 57f
  See also Toluene decomposition
Meat cooking, mixtures of molecules emitted from, 18
Metals
  average concentrations from sediment inside two flooded homes, 175t
  See also Hurricane Katrina aftermath
Mie theory, aerosol radiative properties, 155–156
Modeling
  aerosolized contaminants, 171–172
  Atmospheric Chemical Transport Model (ACTM), 153–154
  book overview, 8–9
  climatic effects of aerosols, 162–164
  film, 172–173
  MUM–Fate model, 82, 84
  overall mass distribution of contaminants, 173, 179, 180t
  reactive fate of polycyclic aromatic hydrocarbons in urban environment, 82, 84
  vapor-phase contaminants, 170–171
  See also Climatic effects of aerosols
Mold
  equation for mold-gas phase partition coefficient, 171
  equations for multi-phase concentrations, 172t
  predicted, film contaminants, 178–179
  predicted aerosolized, spore contaminants, 177–178
  See also Hurricane Katrina aftermath
Monoterpenes
  absorption spectra of, secondary organic aerosol (SOA), 96–98
  chemical composition of, secondary organic aerosol (SOA), 92, 95
  diagram of atmospheric processing of, 92, 94f
  photodissociation spectra of, SOA, 99–101
  secondary organic aerosol (SOA) from atmospheric oxidation of, 92, 95

structures of commonly occurring, 93f
See also Secondary organic aerosol (SOA)
Morphology
oleic acid coatings on particles, 23–24
oleic acid coatings on polystyrene surfaces, 26
oleic acid coatings on silica surfaces, 24, 25f, 26
oleic acid particles and, 27
ozonolysis of oleic acid particles, 14
See also Oleic acid coatings
MUM–Fate model, fate of polycyclic aromatic hydrocarbons in urban setting, 82, 84
Mutagenicity, particulate matter, 34, 36–38
Myristic acid
mixture of oleic acid with, 15, 17
phase diagram, 16f

N

Naphthalene, heterogeneous reaction rate in organic films with gas-phase ozone, 81t
Natural forcing, climate for 2005 vs. 1750, 4, 7f
New Orleans. See Hurricane Katrina aftermath
$(NH_4)_2SO_4$ seed aerosols
categorization of experiments with/without, 121t
choice in aromatic hydrocarbon/NO_x photooxidation system, 112
comparing SOA yields between aromatic hydrocarbons, 116–117
effect of dry, on secondary organic aerosol (SOA) formation, 116–122
experimental, 113–114
hypothesis of dry, effect, 122, 124
smog chamber system, 112, 113f
survey experiments for dry effect of, 115–116
See also Secondary organic aerosol (SOA)
Nitrate aerosols, aerosol source, 3
Nitric acid, deposition from gas phase onto organic films, 85, 86f
Non-thermal plasma (NTP) technology
combining with catalysis, 42–43
corona radical shower, 42
degrading aromatic compounds, 42
spectrum of non–thermal discharge reactor at different voltage, 44, 48f
See also Toluene decomposition

O

Oleic acid coatings
aerosol time of flight mass spectrometry (ATOFMS) kinetic studies, 20
AFM images on silica surfaces for short and long deposition times, 25f
atomic force microscopy (AFM) method, 21
condensing onto dry inorganic substrates, 18–19
contact angle goniometry, 21
deposition on silica by ellipsometry and contact angle goniometry, 25f
ellipsometry method, 21
experimental, 19–21
island formation on silica and PS surfaces, 27
kinetic results, 21–23
morphology, 23–26

particle reactivity and morphology, 27
on particles, 23–24
on polystyrene surfaces, 26
pseudo-first-order rate constants for oleic acid ozonolysis, 22f
PS spin-coated on silica and, deposited on spin-coated PS, 26f
rate constants and oleic acid vapor pressure, 23
sample preparation, 19–20
scanning electron microscopy (SEM) method, 21
SEM of polystyrene latex spheres uncoated and coated with, 24f
on silica surfaces, 24, 26
surface characterization, 21
Oleic acid particles
binary oleic acid mixed systems, 17–18
internally mixed, 15–18
mixtures with dioctyl sebacate, hexadecanoic acid and heptadecanoic acid, 17
multicomponent mixtures of, and other molecules emitted from meat-cooking, 18
oleic acid + myristic acid, 15, 17
oleic acid/n-docosane system, 17–18
phase diagram of oleic acid and myristic acid, 16f
uptake coefficient, 15
See also Oleic acid coatings
Optics, aerosols, 154–156
Organics
in-home, contaminants, 176
measured and predicted concentrations for multiple indoor phases, 177t
Oxidation of monoterpenes. See Secondary , organic aerosol (SOA)
Ozone. See Toluene decomposition
Ozonolysis
oleic acid particles, 14
reaction of ozone with aerosol particles, 18

P

Particles, oleic acid coatings on, 23–24
Particulate matter
air pollution, 32
air quality and climatic seasons in Northern Thailand, 32–33
cell viability and DNA fragmentation studies, 35
Chaing Mai-Lamphun basin in Northern Thailand, 32, 33f
cytotoxicity and apoptosis induction, 37–38
daily levels in Basin, 35, 36f
experimental, 34–35
implications, 38
lung cancer rates in Thailand, 33–34
meteorological contributions, 35
mutagenicity, 36
seasonal averages for two sites in Chaing Mai-Lamphun basin, 35, 37t
toxicity, 36–37
toxicity studies, 34
Perfluorinated compounds
applications, 66
atmospheric implications, 67
compounds in study perfluorododecanoic acid (PFDDA) and partially fluorinated telomer alcohol (8-2 FTOH), 67
experimental vs. calculated footprint of mixtures of stearic acid and lauric acid with PFDDA, 71t
experiments using Langmuir trough, 67–68

isotherms of 8-2 FTOH and stearic acid, 73f
isotherms of 8-2 FTOH collected at surface layer, 71, 72f
isotherms of equimolar samples of stearic acid, lauric acid, and PFDDA, 69f
lauric acid, 68–69
methods of transport, 66
multi component systems of 8-2 FTOH, 73–74
multi component systems of PFDDA, 70–71
PFDDA systems, 68–71
single component systems of 8-2 FTOH, 71–73
single component systems of PFDDA, 68–69
stearic acid, 68
stearic acid addition to 8-2 FTOH, 73f, 74
surfactants, 66–67
systems of 8-2 FTOH, 71–74
Persistent organic pollutant (POP) phenanthrene, 129
See also Phenanthrene, gas phase
Phenanthrene, heterogeneous reaction rate in organic films with gas-phase ozone, 81t
Phenanthrene, gas phase
accumulation of three photooxidation products of, in water film, 136f
bulk and interface air-water partition constants, 134t
comparing reaction rates via radical cations, $^1O_2$ and .OH, 143
concentration change of products during reaction, 136
effect of film thickness on reaction rates, 137, 138f
effect of Suwannee River fulvic acid (SRFA) on reaction rates, 137, 140

equilibrium partitioning at air-water interface, 130–131
equilibrium uptake of, on water films, 132–134
experimental, 129–132
experimental setup, 130
HPLC trace of aqueous film sample after 12 h UV exposure, 135f
materials, 129
measurement of Henry's law constant, 131
measurement of singlet oxygen, 132
observed formation rate constants of 9,10-phenanthrenequinone, 3,4-benzocoumarin, and 9-fluorenone, 143f
observed product formation rate constants in water films, 138t
overall equilibrium concentration in water film, 133
photochemical reactions of, on thin aqueous films, 135–137
photooxidation of, on thin aqueous films, 131
photooxidation on aqueous films containing D_2O, 132
photooxidation pathways of, 140–142
sample analysis, 132
scheme of photooxidation pathways of PAHs in O_2/H_2O system, 141f
SRFA effect on observed formation rate constants of 9,10-phenanthrenequinone, 3,4-benzocoumarin, and 9-fluorenone, 139f
uptake from gas phase on aqueous films with varying thicknesses, 134f
9,10-Phenanthrenequinone (PHEQ)
effect of Suwannee River fulvic acid (SRFA) on formation rate constants, 137, 139f, 140

observed formation rate constants, 141, 143f
product of phenanthrene photooxidation, 135–137
See also Phenanthrene, gas phase
Photochemistry. *See* Secondary organic aerosol (SOA)
Photodissociation action spectrum, monoterpene secondary organic aerosol (SOA), 100
Photodissociation pathways, secondary ozonides and carbonyls, 101, 102f
Photolysis, sunlight and, of organic compounds, 85
Photooxidation
  pathways for phenanthrene, 140–142
  See also Phenanthrene, gas phase; Secondary organic aerosol (SOA)
Pollution
  aerosols, 4
  particulate matter and air, 32
  See also Particulate matter
Polycyclic aromatic hydrocarbons (PAHs)
  air-water interface, 128–129
  atmospheric, 128
  chemical reactions in urban films, 80–84
  environmental pollutants, 128
  heterogeneous reaction rates in organic films with gas-phase ozone, 81t
  homogeneous reactions of gas-phase, with atmospheric oxidants, 128
  modeling rate of, in urban setting, 82, 84
  See also Atmospheric chemistry; Phenanthrene, gas phase
Polystyrene
  oleic acid deposited on spin-coated PS, 26
  spin-coated on silica, 26
  See also Oleic acid coatings
Primary biogenic aerosols, aerosol source, 3
Pyrene, heterogeneous reaction rate in organic films with gas-phase ozone, 81t

R

Radiative forcing, soil dust, 164
Radiative forcing of climate
  aerosols and climate, 159–160
  anthropogenic and natural, for 2005 vs. 1750, 4, 7f
  atmospheric aerosols, 3–4
  diagram of radiative mechanisms, 5f
  effect of aerosols, 1
  See also Climatic effects of aerosols
Radiative transfer, aerosols, 156–157
Reactivity
  species in particles, 14
  See also Oleic acid coatings; Oleic acid particles

S

Salmonella typhimurium, mutagenicity and particulate matter, 34, 36
Scanning electron microscopy (SEM)
  method, 21
  polystyrene latex spheres uncoated and coated with oleic acid, 23, 24f
Sea salt, aerosol source, 3
Seasons, variations in particulate matter, 35, 37t
Secondary organic aerosol (SOA)
  absorption spectra of monoterpene SOA, 96–98

absorption spectrum of organic material from ozonolysis of D-limonene, 97*f*
aerosol photochemistry, 96
approach for studying SOA photochemistry, 99*f*
atmospheric implications, 124–125
atmospheric oxidation of monoterpenes, 92, 95
calculation of SOA yields, 114–115
categorization for experiments with/without $(NH_4)_2SO_4$ seed aerosols, 121*t*
chemical composition of monoterpene SOA, 92, 95
commonly occurring monoterpenes, 93*f*
comparing absorption spectrum and two photodissociation action spectra for D-limonene SOA, 100*f*
comparing SOA yields between aromatic hydrocarbons, 116–117
diagram of atmospheric processing of monoterpenes and oxidation products, 92, 94*f*
effective extinction coefficient estimation, 98
effect of dry ammonium sulfate seeds on SOA formation, 116–122
equation for yield, 114
experimental, 113–114
formation, 91
future directions, 101, 103
hypothesis of dry $(NH_4)_2SO_4$ seeds effect, 122, 124
particle wall loss correction, 115
photochemical aging of organic aerosol, 95–96
photodissociation action spectrum, 100
photodissociation spectra of monoterpene SOA, 99–101
possible photodissociation pathways for secondary ozonides and carbonyls, 101, 102*f*
proposed oligomerization processes, 124*t*
schematic of smog chamber, 113*f*
smog chambers for studying formation, 112
survey experiments for dry ammonium sulfate seed effect, 115–116
toluene/NO_x photooxidation systems, 117, 121*t*, 122
variations of NO_x–NO, O_3 and PM concentration in survey experiments of toluene, m-xylene, and 1,2,4-trimethylbenzene, 118*f*, 119*f*
m-xylene/NO_x and 1,2,4-trimethylbenzene/NO_x photooxidation systems, 122, 123*f*
yield and empirical SOA formation model, 114–115
yields variation with organic aerosol mass concentration for toluene/NO_x photooxidation experiments, 120*f*
yield variation with organic aerosol mass concentration for seeds-free aromatic hydrocarbons/NO_x photooxidation experiments, 120*f*
Secondary ozonides, photodissociation pathways, 101, 102*f*
Sediment
experimental analysis, 170
geochemistry, 173–175
in-home, contaminants, 175–176
metals, 175–176
organics, 176, 177*t*
See also Hurricane Katrina aftermath

Silica surfaces, oleic acid coatings on, 24, 25f, 26
Singlet oxygen
  decay rate, 141
  measurement, 132
Smog chambers, studying secondary organic aerosol (SOA) formation, 112
Soil dust
  aerosol source, 2
  atmospheric circulation, 164
Stearic acid
  addition to fluorotelomer alcohol, 73–74
  experimental and calculated footprint of mixture of, with perfluorododecanoic acid (PFDDA), 71t
  isotherms of fluorotelomer alcohol and, 73f
  mixture with PFDDA, 70–71
  surface pressure-area isotherm, 68, 69f
  See also Perfluorinated compounds
Sulphate aerosols, aerosol source, 3
Surface activity. See Perfluorinated compounds
Surfaces, reflectivity of underlying, 150
Surfactants, perfluorinated compounds as, 66–67
Suwannee River fulvic acid (SRFA)
  effect on phenanthrene photooxidation rates, 137, 139f, 140
  See also Phenanthrene, gas phase

T

Thailand. See Particulate matter
Toluene decomposition
  apparatus and analysis procedures, 56, 58
  degradation rate and enhancement factor, 58, 60
  degradation route by Fe^{2+} catalysis in corona radical shower, 51f
  effect of catalyst concentration on, 53
  enhance factor of toluene degradation rate vs. Fe^{2+} concentration, 55f
  homogeneous catalytic reaction induced by transition metal ions, 56, 57f
  homogeneous enhancement of, with Fe^{2+}, 50–51, 53
  HPLC analysis of intermediate product of, in liquid, 49f
  laboratory experimental studies, 56, 58–60
  ozone concentration toward voltage without and with added toluene, 46f, 47f
  possible mechanism, 50, 51f
  preparation of metal ions solution, 58
  reactions in droplets, 44–45, 50
  reactions in gas phase, 43–44
  schematic diagram of experimental system, 59f
  spectrum of non-thermal discharge reactor at different voltage, 48f
  toluene average degradation rate vs Fe^{2+} concentration, 54f
  total oxidant detection and, at different voltages, 52f
Toxicity, particulate matter, 34, 36–38
Transition metal ions, homogeneous catalytic reactions, 56, 57f
Twomey effect, indirect effect of aerosols, 150–151, 159–160

U

Ultraviolet (UV) photooxidation. See Phenanthrene, gas phase

Urban films
  chemical reactions on, 80–84
  nature of, 80
  photochemistry of compounds associated with, 85–87
  See also Atmospheric chemistry

V

Vanadium, average concentrations from sediment inside two flooded homes, 175t
Vapor-phase contaminants
  equations for multi-phase concentrations, 172t
  modeling, 170–171
  predicted, 176–177
  See also Hurricane Katrina aftermath
Visibility, aerosols, 4
Volatile organic compounds (VOC)
  atmosphere, 112
  emission, 42
  non-thermal plasma (NTP) technology, 42
  See also Toluene decomposition
Volcanoes, aerosol source, 3

W

Water films, atmospheric
  effect of film thickness on phenanthrene photooxidation rates, 137, 138f
  equilibrium uptake of gas-phase phenanthrene, 132–134
  photochemical reactions of phenanthrene, 135–137
  photooxidation of phenanthrene, 131
  See also Phenanthrene, gas phase